Introduction to Contextual Maths in Chemistry

Chemistry Student Guides

Editor-in-Chief:
Julie Macpherson, *University of Warwick, UK*

Series editors:
Dudley Shallcross, *University of Bristol, UK*
Paul Taylor, *University of Leeds, UK*
Simon Humphrey, *University of Texas at Austin, USA*

Addressing a challenging concept in undergraduate chemistry, each book in the *Chemistry Student Guides* series aims for the reader to achieve a 'light bulb' moment. Once a concept is understood, the reader will no longer need to use rote memorisation techniques, but instead will have the confidence to see how the learning can be applied across a range of problems, in order to obtain the correct answers.

Each book received student input so it targets exactly where students struggle. Essential student involvement, combined with high editorial engagement from the Board, means that each title is closely tuned to audience needs. The books can be used as a supplement or alternative to core textbooks, and will also work well as a revision tool during exam time.

Titles in the series:

1: Introduction to Stereochemistry
2: Introduction to Contextual Maths in Chemistry

How to obtain future titles on publication:
A standing order plan is available for this series. A standing order will bring delivery of each new volume immediately on publication.

For further information please contact:
Book Sales Department, Royal Society of Chemistry, Thomas Graham House, Science Park, Milton Road, Cambridge, CB4 0WF, UK
Telephone: +44 (0)1223 420066, Fax: +44 (0)1223 420247,
Email: booksales@rsc.org
Visit our website at www.rsc.org/books

Introduction to Contextual Maths in Chemistry

By

Fiona Dickinson
University of Bath, UK
Email: f.dickinson@bath.ac.uk

and

Andrew McKinley
University of Bristol, UK
Email: a.w.mckinley@bristol.ac.uk

ROYAL SOCIETY
OF **CHEMISTRY**

Chemistry Student Guides, No. 2

Print ISBN: 978-1-78801-425-0
EPUB ISBN: 978-1-83916-193-3
Print ISSN: 2632-9867
Electronic ISSN: 2632-9875

A catalogue record for this book is available from the British Library.

The Royal Society of Chemistry is a charity, registered in England and Wales, Number 207890, and a company incorporated in England by Royal Charter (Registered No. RC000524), registered office: Burlington House, Piccadilly, London W1J 0BA, UK, Telephone: +44 (0) 20 7437 8656.

Visit our website at www.rsc.org/books

Printed in the United Kingdom by CPI Group (UK) Ltd, Croydon, CR0 4YY, UK

For Rowan,

Who loves a book.

Te queremos...

"The teaching of mathematics to students of chemistry has always been a difficult undertaking. Many students do not appreciate the value of mathematics in their subject and their confidence is often further undermined by textbooks which rely heavily on examples drawn from physics. Dickinson & McKinley bring a great deal of experience of chemistry teaching at degree level and their textbook is explicitly organised so that mathematical concepts are mapped directly to topics a chemistry student will find familiar. The book also places emphasis on techniques of dimensional analysis and the statistical analysis of data, which are often poorly taught to chemistry students. Overall the book fills a gap in chemical education and is highly recommended to anyone teaching physical chemistry or laboratory skills to chemistry students in higher education."

Ben Horrocks
Newcastle University, UK

"*Introduction into Contextual Maths for Chemistry* highlights and captures the fundamental mathematical concepts that arc used within chemistry. I found that the book teaches maths in a way that is easily digestible to a student. There are no "big words" or fancy maths terminology meaning that if you haven't seen a certain maths concept before, it's easy to understand what is being taught. There are case studies which show how each maths concept works alongside a mhemistry theory which I found was really useful in consolidating the relationship between chemistry and maths. I would definitely recommend this book to any student, whether you have studied maths previously or not, as it teaches the fundamental concepts which are used within a chemistry degree."

Isabel Manuel
Chemistry Student, University of Bath, UK

"The book covers a wide range of topics, giving rounded support for the maths required for a chemistry degree. It is broken down well into manageable chunks with context laced throughout which help to add depth to the concepts being learned or revised. This is useful for people, such as myself, who learn better by applying concepts to 'the bigger picture'. The breakdown of the vocabulary is also really useful, as lecturers often don't realise that they're introducing a new concept, or explain it very briefly, so having a source to consolidate that and ensure I've understood their explanation is very helpful. The increased complexity of the equations I was dealing with compared to A-Level drove consideration of units from my mind, but the way this book breaks down the section on units and gives examples would probably save me from losing easy marks in an exam."

Cerys Day-Williams
Chemistry Student, University of Bath, UK

Author Biographies

Fiona Dickinson is a Lecturer in the Department of Chemistry at the University of Bath, where she teaches courses on maths and physical chemistry. She has long been interested in the application of maths in chemistry, after having undertaken a first degree in chemical physics and subsequent doctoral research at the University of Newcastle upon Tyne. As a Teaching Fellow and Lecturer, she has worked at a number of UK institutions, giving her a perspective on the common problems undergraduates face in applying pure mathematics to chemical problems.

Andrew McKinley is a Lecturer in the School of Physics at the University of Bristol, having previously worked as a Teaching Fellow in the Department of Chemistry at Imperial College London and later at the re-established Department of Chemistry at Swansea University. He has devoted his work to improving students' transition to higher education; through this work, he recognises that a perceived lack of mathematical fluency often poses a barrier to students making the most of their studies of chemistry. Andrew graduated with degrees from Imperial College London and the University of Newcastle upon Tyne and has taught on a range of mathematics, physical chemistry and general physics courses.

Student Contributors

We are grateful for the input of our student focus groups, without whom we would not have had a book focussing on the needs of students. Their contributions have been included in the book and were vital for ensuring that our words – while forming a summary of that which was taught – did not become too 'dry' in the context of a textbook and preserved, as accurately as possible, the learning intentions which we embody in our teaching styles.

There is no order which does the contributions justice, so we have simply listed our contributors alphabetically. Thank you all for your input, it was invaluable!

Henry Apsey	Swansea University
Amman Bhardwaj	University of Warwick
Emily Brogden	University of Warwick
Sarah Brown	University of Bath
Annabel Brunt	University of Bath
Dominic Bussey	Swansea University
Jac Evans	Swansea University
Patrick Francis-Whitehouse	Swansea University
Olivia Henderson	University of Bath
William Knight	Swansea University
Morgan McKee	Swansea University
Albor Meha	Swansea University
Paige Mitchell	Swansea University
Bart Payne	University of Warwick
Sophie Powell	Swansea University
Chris Rayer	Swansea University
Bethany Tack	Leeds University
Matt Taylor	University of Warwick
Joseph Wakefield	University of Bath
Will White	University of Bath
Jack Williams	University of Leeds
Mitchell Wood	Swansea University
Daniel York	Swansea University

Preface

Mathematics is the language by which we describe and model the natural world around us. Mathematics is challenging, sometimes abstract, but it is absolutely logical and is one of the few disciplines where absolute certainty of proof can be realised. To fully understand and explain the chemical phenomena which present themselves during the course of our studies in chemistry, from predicting reaction rates to explaining thermodynamic principles, we must be able to use the tools provided for us by mathematics. We do not always need to understand the full mathematical underpinning of our chemical processes; however, by knowing how the tools were created, we can gain deeper insight into the world inside our reaction flasks.

Mathematics is often seen as a challenging subject; all too often the phrase '*I can't do maths*' is heard in reference to numerical problems before algebraic problems are attempted. The reality is that maths is a language and, just like every language, it takes *practice* to develop a level of fluency. From simple numerical exercises (such as adding up the cost of everything in your weekly shopping basket to avoid nasty surprises at the checkout) to algebraic problems (such as scaling up a recipe which 'serves four' for your party of ten), the opportunities for practising mathematics are everywhere. Through developing this fluency at lower levels, you will gradually increase your confidence in handling higher-order problems.

In chemistry, maths is vital for conceptual understanding, whether we are balancing chemical equations, plotting data and analysing graphs or applying the principles of quantum chemistry in spectroscopy. Because of the often abstract manner in which mathematics is taught before students embark on a course in chemistry, even those with a relatively high level of maths attainment often struggle to relate the abstract mathematical examples from the maths classroom to the contextual problems in chemistry where they are applied.

Many books are available to provide a guide to maths as you study chemistry; in writing this book, we have sought to keep the focus on chemistry, finding examples from our experience of teaching both chemistry and maths in a chemical context to illustrate the mathematics in action. Some abstraction will be necessary at some points for better clarification of the method; however, we endeavour to bring the discussion back to chemistry as soon as possible!

An overarching theme in this text is that *maths is versatile*; rarely is there a single use for a given mathematical technique in chemistry. While we have tried to show each mathematical concept in isolation in its chemical context, the reality is that problems in chemistry will require several mathematical approaches to solve. This reality has guided the ordering of the content in this book; in the earlier chapters, the mathematical concepts can be applied directly in a chemical context, while in later chapters the chemical context will use new concepts as well as drawing on mathematical techniques from earlier chapters. Through your own application of the techniques in your chemistry study, you will become more fluent in the application of the mathematics to understanding a wider range of chemical phenomena.

What This Book Is For

The mathematics in this book has been considered in a 'chemistry first' approach, with every technique being paired with a chemical context in which it is useful.

It is not our intention for this text to be a definitive mathematical text for the entirety of an undergraduate chemistry programme; rather, the scope of mathematics covered in the book will be limited to those techniques required to support core chemistry material from the first two years of chemical science courses within UK universities. As well as including maths supporting the core chemistry content, we include sections on error analysis and statistics to support laboratory mathematics. Increasingly, these techniques are expected by a range of accrediting bodies, as well as providing a valuable graduate attribute in an era of 'big data'. The study of complex data sets to determine whether differences are 'statistically significant' is an important consideration when scaling reactions to industrial scales or interpreting the impact of new pharmaceuticals, as well as in careers outwith the chemical sciences.

In the preparation of this text, we have consulted student focus groups at the University of Bath and Swansea University, as well as having input from students at the University of Leeds and the University of Warwick, and the text was very much prepared with their guidance. Their thoughts and comments have been embedded through the text as 'tips', in addition to the 'hints' provided at regular intervals. This guidance by students is a central theme in this series, and has given this text a 'student-led' underpinning, ensuring that the mathematical content remains accessible and relevant to the needs of students of the chemical sciences.

Overview of Core Chemistry Topics

Our case studies will generally focus on applications within physical chemistry, as this is the area where mathematics is most prevalent. That said, the organic and inorganic branches of chemistry will also benefit from our discussions when considering such examples as molecular structures and equilibria. The following presents an overview of areas of chemistry and the mathematics which may be required when analysing such problems.

Atomic, Molecular and Crystal Structures

The structures of molecules and their molecular properties (dipole moment, centre of mass) can be modelled with an understanding of the fundamentals of geometry. Trigonometric functions (sine, cosine and tangent functions) are widely used in crystallography, while vector arithmetic can greatly simplify the process of modelling crystal structures. Additionally, when visualising molecular systems, an understanding of different coordinate systems (such as Cartesian and polar coordinates), and when to use each, allows complex systems to be rendered with greater ease.

Laboratory Measurement and Data Handling

Chemistry is a practical subject in which we make our predictions based on prior observations. Therefore, it is vital that we are able to make appropriate measurements and analyse the data appropriately. We will focus on the principles of the uncertainty of measurement as well as the appropriate graphing of data. It is rare in chemistry to simply plot one variable against another; often, we must make a transformation of one or other variable in order to test a series of results against a theory. Therefore, understanding how to rearrange an equation to create a linear relationship is a key experimental skill.

Once a value is obtained from our measurements, *via* graphing or other means, we then need to determine the confidence we have in that value. Applying principles of error analysis allows us to ensure that we are reporting our value to the correct precision (rather than defaulting to 'three significant figures'), as well as quantifying the error in that measurement and what we mean when we say 'within the bounds of experimental error'.

Chemical Thermodynamics

Thermodynamics is an exceptionally wide-ranging subject and can be considered one of the 'pillars' of physical chemistry. It draws on a wide range of phenomena, all of which are inter-linked and can be used to determine the overall energy of a system. For example, the humble gas law, $pV = nRT$, has four interlinked variables to be considered when analysing the energies of such systems.

When determining energy changes and identifying the connection between thermodynamic state functions, we apply *calculus* (the mathematics of change) in addition to our skill in mathematical rearrangements and substitution.

Chemical Kinetics

While thermodynamics examines whether chemical processes are favourable or otherwise, kinetics looks at *how fast* these processes happen. As with thermodynamics, the rate of a reaction is governed by a range of phenomena, including concentrations, temperature, diffusion rates and so on. Our models of chemical kinetics are centred around the mathematical principles of growth and decay, namely exponentials and logarithms. The Arrhenius relation, $k = Ae^{-E_a/RT}$, is an embodiment of this, showing how the rate constant of a reaction is related to its activation energy and the temperature of the system. We also use the principles of calculus as we determine rate laws for chemical systems and identify the orders of reaction with respect to our reagents; a crucial step in identifying the rate-determining step of a reaction mechanism.

Quantum Chemistry

Quantum chemistry is one of the more abstract areas of chemistry and uses many of the mathematical tools we have already introduced. When thinking about 'wave properties of matter', we use the wave properties of the sine and cosine functions, while the application of polar coordinate systems can often help to simplify our quantum mechanical models of atoms and molecules. Complex numbers in exponential form can simplify wave function arithmetic, and energies of quantum systems can be calculated from this using the principles of differentiation and integration. As a result of these wide-ranging techniques, quantum chemistry is often seen as one of the most complex aspects of chemistry, but is readily managed with understanding of the mathematical techniques. Many aspects of the maths introduced in this text are used in the capstone chapter on 'complex numbers', complementing each other in realising one of the most fundamental—and successful—models in modern chemistry.

How to Use This Book

This book can be used as a single 'sit down and study' mathematics course for first-year students in the chemical sciences; however, it is also intended to be an on-hand reference for the elementary mathematics required for the chemical sciences. We have already mentioned the progression of mathematical techniques in this book; to a certain extent, understanding of each chapter will require some knowledge of preceding chapters, but any such dependencies are linked in the text. As a result, the book has a narrative running through it; however, our intention is that you can 'dip in' to the areas which are relevant to your study. Any previous content needed is signposted, so that you may readily revisit the tools needed for a particular problem.

Each chapter contains a number of 'key concept' boxes (❶†); each of these is a mathematical idea which has been summarised as succinctly as possible in order to better serve as a 'quick reference'. The main body of the text, meanwhile, explains the concept in more detail, describing how it works and giving context to the mathematics in chemistry.

Examples show a key concept in context, while also demonstrating one possible route to solve problems or exercises. Other routes to solve the problem may exist which are equally valid. Some of these examples may ask you to work through specific steps yourself, giving regular checkpoints for feedback before you move on.‡

Checkpoint *exercises* (❓§) follow key concepts (or small groups of key concepts). These exercises comprise short questions which will use the preceding key concepts directly in order to develop understanding of those mathematical ideas. In general, you will find the preceding example helpful in solving the exercises provided. In these exercises we have made a deliberate effort to use a large variety of variable terms, rather than simply using x and y. In our experience, some students find it difficult to link the abstract 'x, y' environment of the maths class to the varied symbols used when applying maths in a chemistry context. The answers to these exercises are provided at the end of each chapter.

To get the most from these exercises, we strongly suggest that you work through the first couple of questions in a given exercise, writing your proposed answer to check against our answers (and consequently to check your understanding of the mathematics). If you do not have the correct answer, try to examine your solution for errors and ensure that you can obtain the correct answer before moving on and completing the rest of the exercise. In general, when practising a technique, the questions within a question set will increase in difficulty through the set, giving opportunities to practise the techniques over a range of examples.

Throughout the book, you will find useful hints (•🔑¶) and tips (💬‖) beside the text. These are reminders or summaries of vital points. Where you see a hint beside an example or exercise, it will guide you through one moment in the solution. You can make the exercises a little challenging by hiding the hints from yourself.

---

†Icon exclamation mark by Louis Buck from the Noun Project.
‡This is a technique based on *programmed learning* – you may find it interesting to read up on this educational approach.
§Icon question mark by José Campos from the Noun Project.
¶Icon key by fix project from the Noun Project.
‖Icon comment by Alice Design from the Noun Project.

Finally, at the end of each chapter, we provide a series of contextual problems, applying the mathematical concepts to more involved chemical problems similar to those you may see in the course of your studies in chemistry; whether in your class learning or the laboratory. In later chapters, you will draw on techniques from earlier chapters in order to solve these problems. Once again, the answers to these problems are provided at the end of each chapter; again, we suggest that you work through the answers on paper and commit to an answer before checking the answers provided.

At the end of the book, we have put together a 'mathematical toolkit' of concepts which are not explicitly covered as part of the text but that some readers may find useful. These variously add extra depth to explanations or simply summarise key mathematical concepts on which the main discussion in the book is based. These are intended as supplemental pieces to augment the text rather than a summary of the concepts within the text.

The book has been written and heavily influenced by our experiences of delivering mathematics courses at a number of institutions to students with a wide range of mathematical abilities. However, we have broadened its scope to be mindful of diverse mathematical qualifications, whether A-level, Scottish Higher, Advanced Placement Qualifications or International Baccalaureate, and our intention is that all students of chemistry with or without such qualifications will benefit from the content herein. Additionally, we have taken inspiration from the accreditation requirements of both the Royal Society of Chemistry (RSC) and the American Chemical Society (ACS), identifying mathematical skills of benefit to all graduates of the chemical sciences, regardless of their eventual career destinations.

Acknowledgements

Writing a textbook is no easy task and we are grateful to a large number of people for their contributions, their proofreading and their willingness to simply listen as we have bounced around ideas for what to include in this text. Most prominently, we thank our student contributors; these are listed separately as they have been invaluable in the preparation of this text.

We thank the series editors, particularly Julie Macpherson (University of Warwick) and Dudley Shallcross (University of Bristol), for selecting our proposal and their initial input at the start of this process, as well as for their exceptionally rapid turnaround of feedback when reviewing drafts of the text. Thanks particularly to Michelle Carey, Sylvia Pegg and Noah Tate at the Royal Society of Chemistry for their exceptional understanding of circumstances!

At the University of Bath, we thank Benjamin Morgan for his insightful comments, as well as Paul Raithby for reading the complete book and Laurie Peters and Gareth Price for their enthusiasm and their preliminary review of maths texts in chemistry.

Thanks go to Bernard Mostert, Joel Loveridge and Jenny Stanford at Swansea University, who all read and commented on early drafts of the text to guide us in our initial deliberations as to what the book should contain.

We thank Ian Gould at Imperial College London for discussions on statistical mechanics and who directly helped to inform the statistics and probability content for this book.

Additionally, we thank the classes of chemistry students who have given feedback on our courses over the years, allowing us to develop the educational approaches used in this book.

We also thank our families, who have had to listen to us talk about this text through its nascence and who, more recently, have become additional proofreaders!

Finally, we owe a debt of thanks to Eimer Tuite; it was our research in the Tuite group at Newcastle University in the mid-2000s which brought our authoring team together and inspired our passion for effective education in chemistry.

List of Symbols and Their Uses in Chemistry

This list is not exhaustive, but tries to summarise the symbols used in chemistry problems in this book

Table 1 Common symbols and their usage in chemistry.

	Variable	Unit	Constant	SI prefix
a_0			Bohr radius	
A	Absorbance			
A		ampere		
Å		ångström ($\times 10^{-10}$) m		
c			Speed of light	
c				centi ($\times 10^{-2}$)
C		coulomb		
d				deci ($\times 10^{-1}$)
e			Charge on an electron	
F	Force		Faraday constant	
G	Gibbs free energy			
G				giga ($\times 10^{9}$)
h			Planck constant	
H	Enthalpy			
Hz		hertz		
i			$\sqrt{-1}$	
I	Electrical current; moment of inertia			
J		joule		
k	Rate constant; spring constant			

Table 1 (*continued*)

	Variable	Unit	Constant	SI prefix
k_B			Boltzmann constant[a]	
k				kilo ($\times 10^3$)
K	Equilibrium constant[b]			
K		kelvin		
L		litre (= dm^3)		
m_e			Rest mass of an electron	
m		metre		milli ($\times 10^{-3}$)
mol		mole		
M	can represent unit of concentration,	mol dm−3		mega ($\times 10^6$)
n	Number of moles			
n				nano ($\times 10^{-9}$)
N	Number of particles			
N_A			Avogadro constant	
N		newton		
p	Pressure			
p				pico ($\times 10^{-12}$)
Pa		pascal		
q	Thermodynamic heat			
r	Separation (of two particles)			
R	Resistance		Gas constant	
R_∞			Rydberg constant	
s	Sample standard deviation			
s		second		
S	Entropy			
S		siemens		
t	Time			
T	Temperature			

Table 1   (*continued*)

	Variable	Unit	Constant	SI prefix
V	Volume; voltage			
V		volt		
w	Thermodynamic work			
W		watt		
δ	*Small or infinitesimal change in*			
Δ	*Change in*			
ε	Molar extinction coefficient; relative permittivity			
κ	Conductivity			
λ	Wavelength			
Λ	Molar conductivity			
μ				micro ($\times 10^{-6}$)
ν	Frequency			
ρ	Resistivity			
σ	Conductivity; population standard deviation			
σ_x	Standard error			
τ	Lifetime			
Ψ	Wave function			
Ω	Multiplicity	ohm		

[a]This is sometimes reported simply as k with no subscript.
[b]Sometimes equilibrium constants appear with a subscript letter.

Table of Contents

Learning Points: What We'll Cover

- Notation of functions and inverse functions
- Higher-order operations: powers, roots, exponentials and logarithms
- Identifying inverse functions of higher-order operations
- Strategies for using inverse functions to rearrange equations
- Dimensional analysis: using units to validate equations and in combining quantities
- SI base and derived units
- Scaling factors: both SI and non-SI
- Interconverting units and identifying a scaling factor

Why This Chapter Is Important

- There is a wealth of mathematical notation which we need to understand to apply the maths in a chemical context.
- The models we use in chemistry are fundamentally mathematical; understanding the mathematical relationships allows us to apply these ideas.
- All these models are represented as equations; being able to rearrange and manipulate them allows us to make predictions and to interpret experimental observations.
- Values are meaningless without a unit; having a consistent and systematic approach to the handling of units means that values can be shared and interpreted with no ambiguity, whether using SI units or quantities which are more manageable in a chemistry context, including dm^3 and wavenumbers (cm^{-1}).
- The process of dimensional analysis allows us to determine the appropriate units for our calculated quantities.

The Basics: Mathematical Functions, Rearranging Equations and Handling Units

Chemistry is a physical science; there is a wealth of equations which describe the behaviour of chemical systems. However, very few of these will be in a form which is easily recognisable, or might not have the *subject* we need in order to carry out the desired calculation. Consequently, we must be able to rearrange equations so that we can ensure that we use the calculations needed to test our reactions against an existing model.

To confidently manipulate these equations, we need to identify mathematical operations and functions, as well as how to *undo* a mathematical function. These are the techniques needed to successfully rearrange an equation.

We must also ensure that we are able to handle units—this not only allows us to correctly determine the results of our calculations, but also allows us to check the validity of our equations through a technique known as *dimensional analysis*. While the *SI base units* are well recognised, we rarely work directly with these quantities, often scaling things up or down (nanometres, millimoles, *etc.*) or working with quantities derived from them (volumes, concentrations, wavenumbers, *etc.*), so it is important to be able to interconvert these units. Finally, we must recognise that units are also subject to the mathematical functions we apply to our values and must be considered alongside these values.

> The subject of an equation is the thing you are trying to calculate; it is separated on its own on one side of the equation.

> Forgetting units and badly rearranging equations are all too easy to do—especially in the pressure of an exam. Try and take some extra time to make sure you don't make these simple mistakes!

1.1 Reversible Reactions, Reversible Mathematics

When we carry out a reaction in chemistry, we are most often interested in the product formed

$$2Na(s) + Cl_2(g) \rightarrow 2NaCl(s) \tag{1.1}$$

In eqn (1.1), we are looking at a simple reaction in which two reactants are combined to create the product, sodium chloride. You may have seen internet videos of this reaction (or been lucky enough to see it in person!) and you will observe that it happens very readily with a large release of energy (so readily, in fact, that it burns with an intensely bright orange flame).

However, if we want to generate sodium metal, we need to think about how we reverse this equation. How can we take sodium chloride as a 'reactant' and convert it to sodium metal and chlorine gas (eqn (1.2))?

$$2NaCl(s) \rightarrow 2Na(s) + Cl_2(g) \tag{1.2}$$

As we pursue a path of sustainable chemistry, increasingly we are searching for reactions which can—ideally—be reversed in a sustainable manner to regenerate the starting materials.

In mathematics, instead of 'reactions' transforming reagents into products, we have 'functions', which transform 'inputs' into 'outputs'. These functions exist in pairs; the first function transforms a value into an output (eqn (1.3)), and the second, often referred to as the *inverse function* (eqn (1.4)), can transform the output back into the original value.

$$3 \xrightarrow{\text{forward function}} 9 \tag{1.3}$$

$$9 \xrightarrow{\text{inverse function}} 3 \tag{1.4}$$

In general, an inverse function is defined so that when it is performed on the output of the original function, you should get the same number back. In mathematics, we notate functions such as this as $f(a)$ for the 'forward function', where a is the variable (or $g(a)$, if we need to illustrate a different function), and $f^{-1}(a)$ for the inverse function. We illustrate these complementary functions in eqn (1.5).

$$
\begin{array}{ccccccc}
3 & \xrightarrow{\text{forward function}} & 9 & \xrightarrow{\text{inverse function}} & 3 & & \\
4 & \xrightarrow{\text{add 6}} & 10 & \xrightarrow{\text{subtract 6}} & 4 & & \\
5 & \xrightarrow{\text{multiply by 2, add 1}} & 11 & \xrightarrow{\text{subtract 1, divided by 2}} & 5 & & (1.5) \\
r & \xrightarrow{f(r)} & r' & \xrightarrow{f^{-1}(r)} & r & &
\end{array}
$$

When there is more than a single operation taking place (e.g. with the transformation of the value 5 shown), notice that in the inverse function we reverse the order of the operations as well as using inverse operations.

The prime symbol has a number of uses in mathematics; it is used here to denote a transformed variable.

> **❶ Key Concept 1.1 Mathematical Functions Notation**
>
> You will have seen mathematics notation $f(x)$, read as 'eff of x'. This simply describes how a function (f) changes with the variable x. The choice of letters is arbitrary, and in chemistry you will often see this style of notation but with different letters or symbols. Some common examples are shown in the table.

Notation	Meaning
$V(r)$	The variation of potential energy V with separation, r
$\psi(x)$	The variation of wave function ψ with position x
$S(U, V, N)$	The variation of entropy S with:
	U (internal energy)
	V (volume)
	N (number of molecules)

We can rearrange any mathematical equation by recognising the inverse function for each instruction in our equations. To think about this in a sensible manner, we need to establish the basics of the mathematical notation (the 'language') which we will be using here and throughout the text.

1.2 An Introduction to Mathematical Operations

As has been mentioned, mathematics is simply a language which we use to describe the world around us; as such, it is often helpful to begin explorations by 'translating' the mathematics into English to better understand what it is asking us to do. You will already be familiar with some of its symbolic vocabulary; *e.g.* very simply, $3 + 4$ is asking you to '*add together the values three and four.*'

We readily recognise that some of these operations 'undo' the action of another; for example, + and − undo the action of each other and × and ÷ undo the action of each other. We use this principle to rearrange equations to make a particular value the *subject* of the equation; we will cover this in Section 1.3 once we have examined some higher powers and their inverse functions.

1.2.1   Powers, Roots, Exponentials and Logarithms

After the simple operations of addition, subtraction, multiplication and division, the next set of operations are powers, roots, exponentials and logarithms. Powers and exponentials share the common structure described in Key Concept 1.2.

> ❶ **Key Concept 1.2 Powers and Exponentials**
>
> Powers and exponentials share a common structure:
> $$b^p$$
> In both cases, there is a *base*, *b*, which is raised to a *power*, *p*.
>
> - In a *power function*, the *base* varies while the power is fixed:
> $$f(a) = a^p$$
> $$f(x) = x^2$$
> $$\text{reaction rate} = k[H^+]^3$$
>
> - In an *exponential function*, the *power* varies while the base is fixed:
> $$g(a) = b^a$$
> $$g(x) = 2^x$$
> $$k(T) = 8^T$$
>
> A power function and an exponential each have their own inverse function, namely a *root* and a *logarithm*, respectively.

1.2.2 Power Functions

You will see power functions appearing throughout chemistry, including expressions of intermolecular forces (for example, the coulombic force between ions is described by $F \propto 1/r^2$; force is inversely proportional to the square of the distance, r, between two ions), orders of reactions (a rate law can be described as $d[A]/dt = k[A]^2[B]^1$, where the overall order is the sum of the powers) and reaction quotients (a reaction quotient can be represented as $Q = [\text{products}]^b/[\text{reactants}]^a$. At its simplest, a power is a representation of repeated multiplication. It is the next step in mathematical operations, in which multiplication can be thought of as repeated addition. This is shown in eqn (1.6)

$$3 \times 4 = 3 + 3 + 3 + 3$$
$$5n = \underbrace{5 + 5 + \ldots + 5}_{n}$$
$$3^4 = 3 \times 3 \times 3 \times 3$$
$$5^n = \underbrace{5 \times 5 \times \ldots \times 5}_{n}$$
(1.6)

By thinking of powers in this way, it becomes possible to see how powers of a common base can combine (eqn (1.7)).

$$3^2 \times 3^4 = \underbrace{3\times3}_{2}\times\underbrace{3\times3\times3\times3}_{4}$$
$$= \underbrace{3\times3\times3\times3\times3\times3}_{6}$$
$$= 3^6$$
$$3^a \times 3^b = 3^{a+b}$$
(1.7)

Fundamentally, when we multiply two powers of a common base together ($4^a \times 4^b$), we obtain the result by simply adding the powers together (4^{a+b}); when we divide, we subtract the powers. This leaves open the possibility that we may end up with a negative value as a power. Laying this out as in eqn (1.8) shows us how we can understand these negative powers

> Powers can only be combined if they are multiplied or divided *and* if they are of a common base: $2^3 + 2^6$ cannot be combined, nor can $2^3 \times 3^4$.

$$3^4 \div 3^2 = \frac{3\times3\times3\times3}{3\times3}$$
$$= \frac{3\times3\times\cancel{3}^1\times\cancel{3}^1}{\cancel{3}^1\times\cancel{3}^1}$$
$$= 3\times3 = 3^2$$
$$3^2 \div 3^4 = \frac{3\times3}{3\times3\times3\times3}$$
$$= \frac{\cancel{3}^1\times\cancel{3}^1}{3\times3\times\cancel{3}^1\times\cancel{3}^1}$$
$$= \frac{1}{3\times3}$$
$$= \frac{1}{3^2} \equiv 3^{-2}$$
$$3^a \div 3^b = 3^{a-b}$$
(1.8)

? Exercise 1.1 The Power of Zero

Using the principle shown in eqn (1.8), show that any number raised to the power of zero (*i.e.* n^0) is always equal to 1:

$$3^0 = 1, \ 10^0 = 1, \ n^0 = 1$$

The final example to look at is what happens when we raise a power function *to another power*. Again, if we lay this out as we did for multiplying or dividing power functions, we can see how to handle this operation, and this is demonstrated in eqn (1.9).

$$(3^2)^3 = \underbrace{3\times3}_{2}\times\underbrace{3\times3}_{2}\times\underbrace{3\times3}_{2}$$
$$= \underbrace{3\times3\times3\times3\times3\times3}_{6}$$
$$= 3^6$$
$$(3^a)^b = 3^{ab}$$
(1.9)

The general rules for how powers can be combined is shown in Key Concept 1.3.

❗ Key Concept 1.3 Combining Powers

Combining powers of a common base is a straightforward process, following these principles:

$$m^a \times m^b = m^{a+b}$$
$$p^a \div p^b = p^{a-b}$$
$$(q^a)^b = q^{a \times b}$$

All powers follow these principles, regardless of whether they are whole numbers, fractions or decimals.

1.2.3 Roots: The Inverse Power Function

As we mentioned earlier, every function must have its inverse. If we have an equation of the form $a = b^c$, we need to be able to rearrange it to find either b or c. In the case of a power function, we already know c (the constant index), so we need to find a way to rearrange the equation to find the base, b. This seems fairly straightforward; if $y = x^2$, then we simply find x by taking the square root of both sides. But what about higher powers?

In Section 1.1, we highlighted the need for inverse functions; whatever mathematical action we perform, we need to be able to 'undo it'. This allows us to find our inverse function for a power, and we show this in Example 1.1.

Example 1.1 The Inverse Power Function

Let's start with a simple function, in which our quantity, r, is squared. To undo this, we need the inverse function, $f^{-1}(r)$, which can be identified as the square root function:

$$f(r) = r^2 \quad \text{and} \quad f^{-1}(r) = \sqrt{r}$$

This means that when we 'stack' (or 'chain') our functions together, we should obtain the original value of r:

$$f^{-1}[f(r)] = \sqrt{r^2}$$

This notation simply shows one function acting directly on the output of the other; we determine $f(r)$ first, then we apply $f^{-1}(r)$ to the result of this function.

There is another way to express this result; as we are dealing with a power function, we can look at Key Concept 1.3 for inspiration. Let's start to apply our thinking about indices:

1. Given that indices either add together ($p^a \times p^b = p^{a+b}$) or multiply together (($p^a)^b = p^{a\times b}$), think what happens when we multiply r^1 by r^1:

$$r^1 \times r^1 = r^{(1+1)} = r^2$$

2. Let's now think about what two numbers can be multiplied together to obtain the value r^1:

$$r^a \times r^a = r^{(a+a)} = r^1$$

3. Knowing how indices combine and that the index on both sides of the equality must be the same, this means that:

$$r^{2a} = r^1$$
$$2a = 1$$
$$a = \frac{1}{2}$$

4. This means that the fractional index $1/2$ means 'take the square root', and the inverse function for r^2 becomes:

$$f^{-1}(r) = r^{\frac{1}{2}}$$

This same approach can be used to identify $r^{\frac{1}{3}}$ as the cube root ($\sqrt[3]{r}$), and $r^{\frac{2}{3}}$ as the cube root of r^2, as our 'stacked indices' multiply together (Key Concept 1.3)

$$r^{\frac{2}{3}} = \left(r^2\right)^{\frac{1}{3}} = \left(\sqrt[3]{r^2}\right)$$

⍰Exercise 1.2 Powers: Positive, Negative and Fractional

Using the rules for combining powers, express each of the following as powers in their simplest forms. If they cannot be simplified, identify and explain.

1. $(a^2)^3$

2. $(b^7)^9$

3. $c^3 \times c^8$

4. $d^4 \times c^2 + d^4 \times f^3$

5. $g^{\frac{1}{2}} \times g^{\frac{3}{2}}$

6. $h^{\frac{1}{3}} \div h^{\frac{2}{3}}$

7. $j^{-\frac{2}{7}} \times j^{\frac{9}{7}}$

8. $(k^{\frac{4}{5}})^5$

9. $(m^{\frac{4}{3}})^{-2}$

The number e is a *transcendental number*, like π, with an unending string of random decimal places starting 2.718.... It is known as Euler's number after Leonhard Euler (pronounced 'Oiler'), while the number itself arose from a study about compound interest on financial transactions! A definition of the number is shown in Toolkit A.12, where $e^1 \approx$ 2.718...

Depending on the book, lecturer or even context, the application of the number e may have a notation change and be written as exp(A), to mean the same as e^A. It's always worth checking this whenever you see $\exp(-E_a/RT)$!

1.2.4 Exponential Functions

The exponential function is a mathematical tool to describe *growth* or *decay*. We see this in chemistry when we consider chemical kinetics; whether we are looking at the propagation and branching of radical reactions or how concentration affects rates of reaction (as reactions proceed, concentrations drop and the rate of reaction decays with this). The typical relationship seen is the Arrhenius equation (and we will revisit this throughout the text), showing a negative, or *decaying* exponential: $e^{-E_a/RT}$.

As mentioned before, an exponential initially looks very similar to a power function, however it is important to recognise that we now have a *constant* base with a *variable* power (see Key Concept 1.2). The form of an exponential now takes the form:

$$5^a = \underbrace{5 \times 5 \times ... \times 5}_{a} \tag{1.10}$$

On the face of it, this looks exactly like the power function; however, we now have the variable a as the index. We see the difference in these functions when we plot graphs of these, shown in Figure 1.1. While the positive component of each graph ostensibly looks the same, the negative values of a show a marked difference; namely, the exponential curve never returns a negative value for $g(a)$.

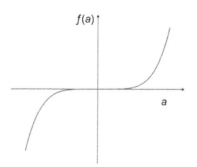
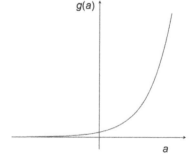

(a) The power function, $f(a) = a^5$.   (b) The exponential function, $g(a) = 5^a$.

Figure 1.1 Comparing a power function with an exponential function. As we have an odd index in the power function (a), the negative values of a return a negative value for the function; however, the exponential function (b) always returns a positive value, whatever the input.

We start to see the difference in the positive values when we plot the graphs on the same axes (Figure 1.2); namely that the growth of an exponential is considerably greater than that of a power function.

The most common exponentials you will see will have a base of either 10 or the number known as Euler's number, e. You will most commonly see exponents of the form 10^a when using scientific notation, while exponents of the form e^a (sometimes represented as 'exp(a)') will most commonly appear when describing chemical reactions, owing to e's origins in describing natural growth.

Having shown the exponential functions, we now need to discuss the inverse operation: the logarithm.

1.2.5   Logarithms: The Inverse Exponential Function

As with the power function, we need to be able to rearrange an equation of the form $a = b^c$; now that we are working with an exponent (constant base, variable index) we already know how to rearrange for b, so now need to rearrange to find the value c. This is less straightforward than for the power functions seen in Section 1.2.3, and to do this, we need to introduce the *logarithm*.

A logarithm is a concept which is often understood poorly, however it is simply the inverse function to an exponential. Put simply, logarithms have the same relationship to exponents as division has to multiplication. Some more detail on the logarithm is provided in Key Concept 1.4.

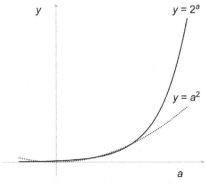

Figure 1.2   Overlaying an exponential function and a power function. Notice that the exponential function (2^a, solid line) increases at a much faster rate than the power function (a^2, dotted line).

> ❗ **Key Concept 1.4 Logarithms and Exponentials**
>
> A logarithm is written in the form
>
> $$y = \log_b(x)$$
>
> where b is the *base* of the logarithm. The logarithm is simply an instruction; it is asking us to find 'the power of b which will give the value of x'. Bases can be any value, but in practical chemistry we only use two logarithms:
>
> - $\log_{10}(x)$, also abbreviated as simply $\log(x)$. This is used in p notation, for pH, pK_a, *etc.* We use the definition $pA = -\log_{10} A$.
> - $\log_e(x)$, more commonly shown as $\ln(x)$. This is the 'natural logarithm', and is found in any natural system with natural decays; *e.g.* radioactive decay, energy state populations, reaction rates, *etc.*
> - These logs may be readily converted through a linear relationship
>
> $$\ln(x) = 2.303 \cdots \times \log_{10}(x)$$
>
> The derivation of this linear relationship is shown in Toolkit A.6.

However, it is enough to simply know the rearrangements, as shown in eqn (1.11).

$$
\begin{aligned}
a &= \log_{10}(b) & \rightarrow \quad 10^a &= b \\
g &= \log_e(h) \equiv \ln(h) & \rightarrow \quad e^g &= h \\
k &= \log_x(m) & \rightarrow \quad x^k &= m
\end{aligned}
\tag{1.11}
$$

> ❗ **Key Concept 1.5 Combining Logarithms**
>
> There are three main properties of logarithms, concerning addition, subtraction and the effect of powers within the logarithm:
>
> - Logarithm addition:
>
> $$\log_x A + \log_x B = \log_x(AB)$$
>
> - Logarithm subtraction:
>
> $$\log_x A - \log_x B = \log_x\left(\frac{A}{B}\right)$$

- Powers inside logarithms (see first example on addition of logarithms to support this):

$$\log_x\left(A^3\right) = \log_x\left(A \times A \times A\right)$$
$$= \log_x\left(AAA\right)$$
$$= \log(A) + \log(A) = \log(A)$$
$$= 3\log_x\left(A\right)$$

- Effect of a negative log (application of powers):

$$-\log_x\left(A\right) \equiv \log_x\left(A^{-1}\right) \equiv \log_x\left(\frac{1}{A}\right)$$
$$-\log_x\left(\frac{A}{B}\right) = \log_x\left(\frac{B}{A}\right)$$

?Exercise 1.3 Interconverting Between Exponentials and Logarithms

Simplify the following logarithms and write them in their exponential form; for example, $\log_b(a) = c$ can be written as $b^c = a$. Where possible, determine the value of the algebraic unknown. For example:

$$\log_2 512 = t$$
$$2^t = 512; t = 9$$

1. $\log_3 a = 9$
2. $\log_5 125 = b$
3. $\log_c 64 = 6$
4. $\log_2 d + \log_2 f = 10$
5. $\log_g x = 4$
6. $\log_6 h = 4$

Use the logarithmic form to write the following equations in terms of the variable index; for example, $a = b^c$ can be written as $\log_b(a) = c$.

7. $a = 3^4$
8. $125 = b^3$
9. $32 = c^5$
10. $d = 7^3$
11. $625 = f^4$
12. $1024 = 2^g$

One way of thinking about a logarithm is to 'translate' it as a question: the expression $\log_{10}(b)$ is asking, 'What power of the base 10 is needed to give us the value of b?'

To see what is happening with the logarithm–exponent pairs in eqn (1.11), just keep track of where the letters are in each relationship. In the first example, a is simply the power to which 10 must be raised in order to obtain the value b.

1.2.6 Logarithms in a Chemical Context

We have introduced the logarithm, as well as some rules for combining logarithms (Key Concept 1.5), however it is useful to see these in context as this gives us a real-world example of these mathematical concepts in action.

Remember that the definition of equilibrium requires that the overall Gibbs energy is zero. However, as equilibria are frequently established at conditions away from standard conditions, we can have a non-zero *standard* Gibbs energy for the reaction.

Case One: Calculating Gibbs Energies

The standard Gibbs energy for a process at equilibrium can be calculated using the relationship shown in eqn (1.12), where K_{eq} is the equilibrium constant for the reaction:

$$\Delta G^{\ominus} = -RT \ln K_{eq} \qquad (1.12)$$

The equilibrium constant for a general reaction may be found from the equilibrium concentrations, as:

$$A + B \rightleftharpoons C + D \qquad (1.13)$$

$$K_{eq} = \frac{[C][D]}{[A][B]}$$

Substituting the result of eqn (1.13) into eqn (1.12), our standard Gibbs energy for the 'forward' process is then:

$$\Delta G^{\ominus}_{forward} = -RT \ln\left(\frac{[C][D]}{[A][B]}\right) \qquad (1.14)$$

What if we consider the 'back' process? We would expect the same Gibbs energy change, but with the opposite sign. Is this borne out in the logarithm rules? Let's start by expanding the logarithm (eqn (1.15)) and then do the same for the back process (eqn (1.16)):

$$\Delta G^{\ominus}_{forward} = -RT \ln\left(\frac{[C][D]}{[A][B]}\right)$$
$$= -RT\left[\ln([C][D]) - \ln([A][B])\right] \qquad (1.15)$$
$$= +RT\left[\ln([A][B]) - \ln([C][D])\right]$$

$$\Delta G^{\ominus}_{back} = -RT \ln\left(\frac{[A][B]}{[C][D]}\right)$$
$$= -RT\left[\ln([A][B]) - \ln([C][D])\right] \qquad (1.16)$$
$$= -\Delta G^{\ominus}_{forward}$$

Sure enough, we see that the rules of logarithms show us what we already know to be true from our existing understanding of equilibrium processes.

Case Two: Statistical Entropy

The Boltzmann relation for statistical entropy, S (eqn (1.17)), connects the total entropy of a system to the multiplicity of the system (W) *via* the Boltzmann constant, k_B:

$$S = k_B \ln W \qquad (1.17)$$

If two systems, X and Y, are combined, the total entropy of the system will add together. Taking this into account with the rules for combining logarithms, we see that there are implications for the overall multiplicity, as:

$$
\begin{aligned}
S_{total} &= S_X + S_Y \\
&= k_B \left(\ln W_X + \ln W_Y \right) \\
&= k_B \ln (W_X \times W_Y)
\end{aligned}
\qquad (1.18)
$$

We see that when two systems are combined, for the entropy to be additive, the multiplicities must be multiplied. Indeed, this is exactly what happens when two systems are combined. We cover further aspects of such systems in Chapter 2.

1.3 Treating Both Sides Fairly: The Nature of Equality

We are now in a position to consider how we might rearrange equations appropriately in order to make any term the subject of the equation. It is helpful to consider a situation in chemistry where we balance and rearrange equations.

When we consider redox equilibria, in particular those involving redox half-cells, we combine half-cells together. Consider the two half–cell reactions:

$$
\begin{aligned}
A&:\ Zn^{2+}(aq) + 2e^- \rightleftharpoons Zn(s) \quad E^{\ominus} = -0.76\,V \\
B&:\ \tfrac{1}{2}Cl_2(g) + e^- \rightleftharpoons Cl^-(aq) \quad E^{\ominus} = +1.36\,V
\end{aligned}
\qquad (1.19)
$$

Recall from your studies of redox equilibria that we need to combine the equations in such a way as to eliminate the electron term while ensuring that the cell potential, E, is positive. Chemically, we want to have the Cl_2 being oxidised and the Zn being reduced. The method for this is logical, and results in:

$$Cl_2(g) - Zn^{2+}(aq) \rightleftharpoons 2Cl^-(aq) - Zn(s) \quad E^{\ominus} = +2.12V \qquad (1.20)$$

Multiply eqn (1.19) part B by two to make the electron terms equal, then subtract eqn (1.19) part A from the new eqn (1.19) part B. Remember that the $E^{\ominus}$ values *do not change*, as these are *standard state values*.

It looks a little odd, as we have 'minus signs' in place; however, the equation balances. To tidy this up, we remember that whatever we do to one side of the equation, we must do to both sides to ensure that the equality is unchanged.

1. Firstly, let's add $Zn^{2+}(aq)$ to the left-hand side to cancel out this negative. This means that we must add the same to the right-hand side.
2. Finally, we add $Zn(s)$ to the right-hand side; accordingly, we must do the same to left-hand side.

This gives us our final equation, as expected:

$$Cl_2(g) + Zn(s) \rightleftharpoons 2Cl^-(aq) + Zn^{2+}(aq) \quad E^{\oplus} = +2.12V \quad (1.21)$$

The rules are exactly the same in mathematics; whatever we do to one side, we must do to both sides, otherwise our equality no longer applies.

1.3.1   The Application of Inverse Functions

Usually, an equation is written algebraically, using letters to indicate variables and constants. This makes the equation considerably easier to rearrange and allows for identification of terms to cancel. Let's first look at a simple equation you will see many times in chemistry: the Gibbs free energy relation:

$$\Delta G = \Delta H - T\Delta S \quad (1.22)$$

It is a simple equation with simple rearrangement strategies (see Example 1.2), however it demonstrates much of what is needed with regard to rearranging an equation.

Many students come with the approach 'change the side, change the sign'; while this might work for simple addition or subtraction, it is less clear how this can work for multiplication, division or higher-order functions like powers and logarithms. For this reason, it is *always* safer to do the same operation to both sides of the equation as you are less likely to make an error.

Example 1.2 Rearranging the Gibbs Equation

Q: From the Gibbs equation shown in eqn (1.22), make the entropic term, ΔS, the subject of the equation.

A: There are a number of steps to follow, each of which must be followed logically.

1. Firstly, let's add '$T\Delta S$' to both sides; this will cancel out the negative $T\Delta S$ on the right, and add it to the left:

$$\Delta G + \underline{T\Delta S} = \Delta H - T\Delta S + \underline{T\Delta S}$$
$$= \Delta H$$

2. Let's now subtract ΔG from both sides to move it from the left to the right:

$$\Delta G + T\Delta S - \underline{\Delta G} = \Delta H - \underline{\Delta G}$$
$$T\Delta S = \Delta H - \underline{\Delta G}$$

3. Finally, we now divide both sides by T to cancel out the T term on the left and bring it to the right-hand side:

$$\frac{T\Delta S}{\underline{T}} = \frac{\Delta H - \Delta G}{\underline{T}}$$

$$\Delta S = \frac{\Delta H - \Delta G}{T}$$

❓Exercise 1.4 Equation Rearrangements

For each of the equations listed, rearrange to make the suggested term the subject of the equation. For added practice, ensure that you can rearrange the equation to make each part of the equation the subject.

1. $pV = nRT$; T

2. $\Delta G = \Delta H - T\Delta S$; T

3. $K = \dfrac{[H^+] \times [OH^-]}{[H_2O]}$;   [OH⁻]

4. $a = 5b + c$; b

5. $d = 3f^2 - g$; f

6. $s = ut + \dfrac{at^2}{2}$; a

7. $a = \dfrac{b}{\sqrt{c}}$; c

8. $A = \dfrac{\mu D^2 F}{t}$; D

9. $S = 2\pi r^2 + 2\pi rh$; h

10. $Q = \left(\dfrac{c+d}{c-d}\right)$; c

1.3.2   The Arrhenius Equation

The Arrhenius equation is an exemplar of higher-order functions, including multiplication and exponents. The most common form of the Arrhenius equation is:

$$k = Ae^{-\left(E_a/RT\right)} \tag{1.23}$$

> 💬 You may see the natural exponential in eqn (1.23) written as $A \exp(-E_a/RT)$; this is an equivalent notation, which some texts use.

We are often tasked with finding the activation energy (E_a) or the 'pre-exponential factor' A having been given experimental data of rate constants and temperature. Rearranging the Arrhenius equation is therefore a matter of finding the appropriate inverse function for the exponent. As we have a natural decay involving e^x, the appropriate inverse function is the natural logarithm, ln (see Section 1.2.5).

Example 1.3 Rearranging the Arrhenius Equation

There are numerous equations with similar form to the Arrhenius equation in chemistry, the Boltzmann distribution being another example. They may all be rearranged in a similar manner following similar steps; however, make sure that you are clear why each action is being done and the order in which it is done.

Using the form of the Arrhenius equation shown in eqn (1.23), we rearrange to make E_a the subject:

1. Firstly, to isolate the exponential, we divide both sides by A to remove this from the right-hand side:

$$\frac{k}{A} = \frac{A\exp\left(-\frac{E_a}{RT}\right)}{A}$$

$$= \exp\left(-\frac{E_a}{RT}\right)$$

2. To remove the exponent from the right-hand side, we take the natural logarithm of both sides:

$$\ln\left(\frac{k}{A}\right) = \ln\left[\exp\left(-\frac{E_a}{RT}\right)\right]$$

$$= -\frac{E_a}{RT}$$

3. We can now multiply both sides by $(-RT)$ in order to isolate the activation energy, E_a:

$$-RT \times \ln\left(\frac{k}{A}\right) = -RT \times \left(-\frac{E_a}{RT}\right)$$

$$= E_a$$

4. We can write this in a more accessible form, as follows:

$$E_a = -RT\ln\left(\frac{k}{A}\right)$$

This rearrangement will be used in Chapter 4 when we explore how to graph such experimental data.

These 'decaying exponential' functions appear often in chemistry so it is essential that we are able to make such rearrangements; whether we are considering reaction kinetics, energy distributions, the radioactive decay of carbon-14 (^{14}C), energy transfer processes in thermodynamics, ... the list goes on.

1.4 Dimensional Analysis

The technique of dimensional analysis is of vital importance in our work as physical scientists, as units tell us everything we need to know about measurements. Principally, a value is meaningless unless it has an appropriate unit associated with it, and your tutors will insist on your inclusion of appropriate units.

The appropriate use of units can be exceptionally powerful. Units in our quantities allow us to check the validity of an equation; they can tell us if and how quantities can combine; they tell us whether we can indeed take a logarithm of the value (or an exponent) and they allow us to understand the value which we obtain at the end of the calculation.

1.4.1 Base Units and Derived Units

Units are divided into two types; the SI base units and units which are derived from these. The list of SI base units is shown in Key Concept 1.6; the simplest way to think of these is that they are the 'elements' of the units. They are seven units which cannot be expressed in any other way; they cannot be simplified further and they do not relate to each other.

> Some measurements intrinsically have no units associated with them—pH, absorbance, some equilibrium constants, *etc.* In this case, the value is simply reported and there is no need to associate a unit with it. Lengths, volumes, concentrations, *etc. must* have their units associated however—otherwise the value is meaningless!

> Units named after people are still represented with a lower-case initial letter.

❶ Key Concept 1.6 SI Base Units

There are seven SI base units, shown in the table below:

Quantity	Quantity symbol[a]	Unit	Unit symbol
Length	L	metre	m
Mass	M	kilogram	kg
Time	T	second	s
Temperature	Θ	kelvin	K
Amount of substance	N	mole	mol
Electrical current	I	ampere	A
Light intensity	J	candela	cd

[a]The quantity symbols are often used in dimensional analysis (checking on units), a technique developed by James Clerk Maxwell (whom you will no doubt come across in other areas of your studies).

Lecturers often issue a challenge to students to see who can name them; it is alarming how many chemistry undergraduates overlook the importance of the mole as an SI base unit!

Conversely, derived units are derived from the combination of one or more of these base units and allow us to completely describe the physical world. An area may be found by multiplying two lengths together (m × m = m^2) and a volume may be found by multiplying three lengths together (m^3), while a concentration may be found by dividing a quantity of material by a volume (mol ÷ m^3 ≡ mol m^{-3}).

---

❶ Key Concept 1.7 SI Derived Units

Derived units are formed by the multiplication or division of either SI base units or other derived units. Some common derived units are listed in the table, together with their formulae and units.

Quantity	Formula	Unit
Speed	distance ÷ time	m s^{-1}
Force	mass × acceleration	kg m s^{-2}
Kinetic energy (joule)	½ × mass × velocity²	kg m^2 s^{-2}
Thermodynamic work (joule)	force × distance	N m ≡ kg m^2 s^{-2}
Pressure (pascal)	force ÷ area	N m^{-2} ≡ kg m^{-1} s^{-2}

There are many examples of derived units; these are simply ones which you will come across early in your chemistry career.

Note: Units should be represented with a space between each term, and with negative indices rather than a solidus (a 'stroke', /). Think of the confusion which could result from a speed represented as ms^{-1} ('per millisecond') rather than m s^{-1} ('metres per second'), or units of J/K/mol rather than J K^{-1} mol^{-1} (Are we dividing 'J' by 'K/ mol'? That would result in J × mol/K, which is not the same unit). See Toolkit A.5 for more guidance on how indices combine.

You will come across many derived units; indeed, unless you are directly working with one of the SI base units, you will work almost exclusively with derived units.

1.4.2 Congruence of Units

The term 'congruence of units' is the formal way to say that units must be matched appropriately across an equation so that the equation makes sense. There are a number of rules to be considered when looking at units; these are summarised in Key Concept 1.8. It is important to recognise how to do this, as it is a vital way to check equations.

It's really helpful in an exam situation to check the units on both sides of the equation to confirm that you are using the correct equation (or even have remembered it correctly).

---

❶ Key Concept 1.8 Applying Congruent Units in Equations

There are three simple rules regarding units for operations in equations:

1. For first-order operations (addition and subtraction), *units must be the same for each quantity*.

   - *E.g.* you cannot add one second to two moles.
   - Seconds can only add to seconds, moles to moles, *etc.*
   - Centimetres cannot be added to metres; either you must convert cm to m to give a result in metres, or convert m to cm to give a result in centimetres.

2. For second-order operations (multiplication and division), *units may be combined according to the same operation as the values.*

   - A concentration is found by dividing the number of moles present by the volume, resulting in the unit $mol\ dm^{-3}$.
   - The work done in expansion is found by multiplying the pressure (p) by the change in volume (V), resulting in an energy in joules (base units $kg\ m^2\ s^{-2}$).

3. For exponents and logarithms (e.g. e^x, $\log(x)$), *units cannot be used*.

   - Where exponents and logarithms appear, the value will already be unitless, *e.g.* in the Arrhenius equation, the term E_a/RT is unitless, while for pH the expression is formally

$$pH = -\log_{10}\left(\frac{[H^+]}{[H^+]^{\ominus}}\right)$$

   - The use of *standard state conditions* to eliminate units in this manner is common in chemistry and is often seen in thermodynamics.

In chemistry, we use the non-SI dm^3 or litre (L) as the unit of volume, as this is more practical in the context of our chemical reactions.

Example 1.4 Congruence in Equations

In an equation, the units must be equivalent on either side of the equation. Consider the gas law:

$$pV = nRT \qquad (1.24)$$

When all quantities are in SI units, we can consider the units of each quantity in the equation.

Quantity	SI unit	SI base units
Pressure, p	Pa	kg m^{-1} s^{-2}
Volume, V	—	m^3
Number of moles, n	—	mol
Temperature, T	—	K

Dividing both sides by nT allows us to gain an expression for the gas constant, R.

$$\frac{pV}{nT} = R \qquad (1.25)$$

By considering the units on the left-hand side, we will obtain the unit for the gas constant, R.

$$\frac{\overbrace{\text{kg m}^{-1}\,\text{s}^{-2}}^{p} \times \overbrace{\text{m}^3}^{V}}{\underbrace{\text{mol}}_{n} \times \underbrace{\text{k}}_{T}} = \underbrace{\text{kg m}^2\,\text{s}^{-2}\,\text{mol}^{-1}\,\text{K}^{-1}}_{R} \equiv \text{J K}^{-1}\,\text{mol}^{-1} \qquad (1.26)$$

As we can see, through consideration of the SI base units, we have identified the presence of the derived unit joule (see Key Concept 1.7), giving us the units we already recognise for the gas constant R.

By checking the units on either side of our equation, we have a useful tool to validate our equations; if the units do not match up, then we have made a mistake with our values somewhere.

❓Exercise 1.5 Determination of Units from Calculations

For each of the following relationships, determine the base SI units. You will need to start at the beginning, as some early results are used in later questions. The table gives you some units to start.

Symbol	Quantity	SI base units
d or l	distance or length	m
A	area	m^2
V	volume	m^3
v	velocity	m s^{-1}
a	acceleration	m s^{-2}

💬 'I found it so much easier to report the unit in my final answer when I always set the unit as a key part of calculations throughout; it meant I wasn't guessing what the final unit might be. It also showed me that, at times in a calculation, units may cancel out directly, meaning a time-consuming unit conversion wasn't needed.'

1. $F = ma$;   force

2. $p = \dfrac{F}{A}$;   pressure

3. $E = \dfrac{1}{2}mv^2$;   energy

4. $E = Fd$;   energy

5. $C_{p,m} = \dfrac{E}{n\Delta T}$;   molar heat capacity

6. $[A] = \dfrac{n}{V}$;   concentration

7. $Q = It$;   current

8. $V = \dfrac{E}{Q}$;   voltage

9. $R = \dfrac{V}{I}$;   resistance

10. $\rho = R\dfrac{A}{l}$;   resistivity

11. $\kappa = \dfrac{1}{\rho}$;   conductivity

12. $\Lambda_m = \dfrac{\kappa}{[A]}$;   molar conductivity

> Your units here (Question 6) will be in SI, but you probably think about them with a prefix! See Section 1.4.3.

> It is often the case that the same letter is used to represent more than one variable; in question 1.5.8 V stands for voltage, whereas previously it had also been used as the variable for volume. Situations like this appear frequently, as there are only so many symbols available.

1.4.3   Scaling Factors; SI and Non-SI

The last thing we need to do with our units is to cover the various scaling factors which you will come across. The simplest of these scaling factors are the standard SI scaling prefixes, telling us the *order of magnitude* of our measurement. These are the scaling factors seen in millimoles ($\times 10^{-3}$ mol), microseconds ($\times 10^{-6}$ s) or nanometres ($\times 10^{-9}$ m). You will already be familiar with most of these in other areas too (megabits, gigabytes, *etc.*), and these prefixes are listed in Key Concept 1.9. Remember that these are all subject to the rules laid out in Key Concept 1.8; we cannot add moles to millimoles until we have scaled either value appropriately.

---

❶ Key Concept 1.9 Standard SI Scaling Factors

The majority of scaling factors you will come across will be the standard scaling factors shown here, which simply give an easy way to express a magnitude on a value.

Prefix	Symbol	Scaling factor
zepto-	z	$\times 10^{-21}$
atto-	a	$\times 10^{-18}$
femto-	f	$\times 10^{-15}$
pico-	p	$\times 10^{-12}$
nano-	n	$\times 10^{-9}$
micro-	μ	$\times 10^{-6}$
milli-	m	$\times 10^{-3}$
centi-	c	$\times 10^{-2}$
deci-	d	$\times 10^{-1}$
kilo-	k	$\times 10^{3}$

Prefix	Symbol	Scaling factor
mega-	M	$\times 10^6$
giga-	G	$\times 10^9$
tera-	T	$\times 10^{12}$
peta-	P	$\times 10^{15}$
exa-	E	$\times 10^{18}$
zetta-	Z	$\times 10^{21}$

Some things to note about this list:

- With the exception of the kilo-prefix, *positive* powers of ten shown in this list are described by UPPER-CASE prefixes. Deca- (da, $\times 10^1$) and hecto- (h, $\times 10^2$) are also positive powers of ten with a lower-case prefix, however these are very rarely used in chemistry and are not included in this list.
- With the exception of centi- and deci-, all common prefixes differ by 10^3.

Finally there are units which have irregular scaling factors. These are units which often have historical significance or allow us to work with easy values. A typical example is the use of 'atmosphere' as a measure of pressure: 1 atm has a pressure of 101 325 Pa and, as it is the pressure under which we do most of our bench chemistry, it is a convenient scaling factor to work with. However, we must take this into account in calculations; we cannot add a pressure in atmospheres to a pressure in pascals, and must either interconvert or check to see whether the units will cancel. Examples of such irregular scaling factors are listed in Key Concept 1.10; however this list is not exhaustive.

❶ Key Concept 1.10 Units with Irregular Scaling Factors

This is a list of some units with irregular scaling factors. They are provided with the scaling factors to SI units and the situation in which you would expect to see them.

Unit	Symbol	Application	SI scaling factor
ångström	Å	bond length	1.000×10^{-10} m
litre	dm^3	volume	1.000×10^{-3} m^3
wavenumber	cm^{-1}	energy	1.986×10^{-23} J
electron-volt	eV	energy	1.602×10^{-19} J
atmosphere	atm	pressure	1.013×10^5 Pa
bar	bar	pressure	1.000×10^5 Pa
torr	mm Hg	pressure	1.333×10^2 Pa
dalton	Da	molecular mass	1.000×10^0 g mol^{-1}

1.5   Interconverting Units

Our final section in this chapter concerns the interconversion of units. As has been said throughout the chapter, we can only add units which are congruent (*e.g.* adding centimetres to centimetres), and when multiplying or dividing units, it is helpful if quantities such as volumes are all converted to the same unit (*i.e.* convert all volumes in the equation to dm^3, rather than dividing m^3 by dm^3).

A key question in interconverting units is, '*Am I expecting the value to get bigger or smaller?*' When interconverting, we will either be multiplying or dividing by the scaling factor. The easiest way to demonstrate this process is by example. Example 1.5 gives a focus on converting a length in cm, as well as a vibrational frequency in cm^{-1}.

> It can be confusing, particularly with inverse quantities; for example, converting wavenumbers in cm^{-1} ('per centimetre') to m^{-1} ('per metre'). If you have '50 in a centimetre', you would expect to have one hundred times as many in a metre, so 50 cm^{-1} becomes 5000 m^{-1}.

Example 1.5 Identifying and Applying a Scaling Factor

1: Convert 168 cm into metres.

- Looking at standard tables (see Key Concept 1.9), we identify that the centi-prefix 'translates' to $\times 10^{-2}$. We can then replace the 'cm' with '$\times 10^{-2}$ m':

$$168 \text{ cm} = 168 \times 10^{-2} \text{ m}$$

- Now we apply the power of ten appropriately to give our result:

$$168 \times 10^{-2} \text{ m} = 1.68 \text{ m}$$

2: Convert the infrared frequency 1760 cm^{-1} into SI units.

- We are now working with a reciprocal unit. We already know our scaling factor from before, but it might be suitable to write this mathematically:

$$1 \text{ cm} = 1 \times 10^{-2} \text{ m}$$

- We know we need to work in reciprocal units ('$^1/_{unit}$'), so let's find the reciprocal of both sides:

$$\frac{1}{1 \text{ cm}} = \frac{1}{1 \times 10^{-2} \text{ m}}$$
$$1 \text{ cm}^{-1} = 1 \times 10^2 \text{ m}^{-1}$$

- Now we replace the term 'cm^{-1}' with '$\times 10^2 \text{ m}^{-1}$' in our original infrared frequency and express the result in scientific notation:

$$1760 \text{ cm}^{-1} = 1760 \times 10^2 \text{ m}^{-1} = 1.760 \times 10^5 \text{ m}^{-1}$$

- We see here that the number has become larger; however, if we reason it through, this makes sense if we think of the cm^{-1} to mean 'per cm'—there are 1760 'per centimetre', so there must be one hundred times as many 'per metre'!

Another complication comes when we have reciprocal units; we examine the wavenumber (cm^{-1}) in Example 1.5, but we also routinely come across reciprocal units when considering concentration. This is a more complex example as it is a compound unit (*i.e.* it has a molar quantity combined with a volume). We show the handling of these compound units in Example 1.6

Example 1.6 Converting Compound Units

A solution is prepared to a concentration of 2.8 μmol cm^{-3}. Determine the concentration in mmol dm^{-3}.

- We need to think about a number of conversion factors. We refer to the tables in Key Concepts 1.9 and 1.10 to identify the terms we need; in considering the concentration, we need to dig out a few factors:

  – One micromole (μmol) is equal to 1×10^{-6} mol.
  – One millimole (mmol) is equal to 1×10^{-3} mol.

- Expressing this mathematically will allow some manipulation:

$$1 \text{ μmol} = 1 \times 10^{-6} \text{ mol}$$

$$1 \text{ mmol} = 1 \times 10^{-3} \text{ mol}$$

- We now rearrange each of these to make the 'mol' term the subject (*i.e.* to collect all the scaling factors on one side)

$$A: \quad \frac{1 \text{ μmol}}{1 \times 10^{-6}} \equiv 1 \times 10^{6} \text{ μmol} = 1 \text{ mol}$$

$$B: \quad \frac{1 \text{ mmol}}{1 \times 10^{-3}} \equiv 1 \times 10^{3} \text{ mmol} = 1 \text{ mol}$$

- We can now equate the two equations:

$$1 \times 10^{6} \text{ μmol} = 1 \text{ mol} = 1 \times 10^{3} \text{ mmol}$$

and rearrange to make μmol the subject:

$$1 \text{ μmol} = 1 \times 10^{-3} \text{ mmol}$$

This gives us our conversion factor. If we wish to convert our value into mmol, we must multiply our value by the conversion factor 1×10^{-3}. This makes sense; a concentration of 2 μmol will be smaller than 2 mmol, so when converting we would expect a smaller value.

- We now turn our attention to the volume. We can follow the same process as before to identify that $1 \text{ cm} = 1 \times 10^{-1}$ dm; however, we are dealing with cubic centimetres, not linear centimetres. Consequently, we must cube this equation:

$$1 \text{ cm} = 1 \times 10^{-1} \text{ dm}$$

It is easy to call a 'dm^3' a 'decimetre', rather than a 'decimetre cubed'. This is a frequent source of mistakes in unit conversions!

$$(1\,cm)^3 = (1 \times 10^{-1}\,dm)^3$$

$$1\,cm^3 = 1 \times 10^{-3}\,dm^3$$

- This gives our conversion factor of 1×10^{-3}. However, there is one last thing to observe. Our concentration is moles per cubic centimetre (*i.e.* mol cm^{-3}). Therefore, we must take the reciprocal on both sides to identify cm^{-3}:

$$1\,cm^{-3} = 1 \times 10^3\,dm^{-3}$$

- Therefore our scaling factor will make the value larger by a factor of 10^3. This makes sense; if we have 2.8 μmol per cubic centimetre, we would expect to have more per cubic decimetre, as there are 1000 cm^3 in 1 dm^3, and each cubic centimetre contributes 2.8 μmol to the whole.
- Therefore, our final answer is first multiplied by 10^{-3} to convert our quantity from μmol to mmol, and it is then multiplied by 10^3 to convert our volume scalar from 'per cubic centimetre' to 'per cubic decimetre'. These two factors cancel out, and we end up with a final concentration of 2.8 mmol dm^3.

? Exercise 1.6 Unit Conversions

Convert the following units:

$N_A = 6.02214 \times 10^{23}$ mol^{-1}

1. 3.4 μm to nm and m
2. 22.4 dm^3 to m^3 and cm^3
3. 270.4 g mol^{-1} to kg mol^{-1} and zg molecule^{-1}
4. 3.4 min^{-1} to s^{-1} and h^{-1}
5. 3400 cm^{-1} to μm^{-1} and m^{-1}
6. 3.79 mol^{-1} m^3 min^{-1} to mol^{-1} dm^3 s^{-1}

1.6 Summary

In this chapter, we have covered a number of the basics needed to carry out a course of study in chemistry. The notation used in mathematics is often a source of confusion to those unfamiliar with it; but remember that it is simply a language, telling you an instruction to carry out. It is often worth thinking, 'What are these symbols asking me to do?'; as we go through this text, we will explain the mathematical symbols in terms of 'plain English' instructions.

Handling units is a vital practical skill; any value we present in our chemical problems is meaningless without units (unless, through the course of mathematical manipulations, the units cancel out!), and, as such, values should always be presented with their appropriate units. Alongside units are the standard scaling factors;

these are used to allow us to express values in 'sensible' terms; it is much easier to express a fluorescence lifetime as 22 ns than as '0.000000022 s' or even 2.2×10^{-8} s; however, we must remember that all of these are mathematically correct.

Summary: Chapter 1

Mathematical Functions

- The notation for a function is in the form $f(x)$, where f is a function which changes with variable x. The terms 'f' or 'x' are placeholders and other letters may be used to suit the context.
- Two functions ($f(a)$ and $g(a)$) can be combined in two ways:

  - The outputs of each function can be multiplied together, notated $f(a) \times g(a)$.
  - The output of one function can be fed directly into a second function, notated $f[g(a)]$, called a 'chain'. The inner function is calculated first, and the outer function acts on the result of this.
  - The order of the chain is important: in the vast majority of cases, $f[g(a)] \neq g[f(a)]$.

- Every mathematical function, $f(x)$, will have its inverse function, $f^{-1}(x)$, such that $f^{-1}[f(x)] = x$. This inverse function will 'undo' the original function to return the initial value.

Powers, Exponentials and Logarithms

- A power function has a variable base and a constant index; e.g. x^2, r^{-6}, $a^{1/4}$.
- Powers of a common base can be represented as

$$a^{-p} \equiv \frac{1}{a^p}$$

$$a^m \times a^p = a^{(m+p)}$$

$$\frac{a^m}{a^p} = a^{(m-p)}$$

$$(a^m)^p = a^{(m \times p)}$$

$$a^{1/2} \times a^{1/2} = a^1 \, ; \, a^{1/2} = \sqrt{a}$$

- An exponential function has a constant base and a variable index; e.g. 2^x, 6^{-r}, $4^{a/4}$.
- Exponents of a common base can be combined as for power functions.
- A logarithm is the inverse function of an exponential function with the same base, *i.e.* a logarithm will 'undo' an exponential.

- A logarithm of a value c will tell you the power of the base needed to obtain the value of c.
- Logarithms of a common base can be combined using the 'rules of logarithms':

$$\log_x A + \log_x B = \log_x (AB)$$

$$\log_x A - \log_x B = \log_x \left(\frac{A}{B}\right)$$

$$\log_x (A^3) = 3\log_x (A)$$

$$-\log_x \left(\frac{A}{B}\right) = \log_x \left(\frac{B}{A}\right)$$

- The 'natural logarithm', $\ln(A)$ can be written as $\log_e(A)$.

Units and Dimensional Analysis

- Quantities can only be added or subtracted if their units are the same.
- When quantities are multiplied, divided or raised to a power, their units are also multiplied, divided or raised to the power, in line with the values.
- Exponents and logarithms can only operate on values and cannot operate on units.
- Units must be the same on both sides of an equation.
- Some quantities are unitless; these are simply reported without units, e.g. pH = 4.2.

❓Exercise 1.7 Handling Units and Rearranging Equations in Context

---

1. The famous Einstein equation $E = mc^2$ is more properly written $E^2 = p^2 c^2 + m_0^2 c^4$. Determine the units of the variable p, where m and m_0 are masses, and c is the speed of light.
2. The entropy change during expansion of a gas is given by eqn (1.27). Determine the ratio of the initial and final volumes for expansion of exactly 2.5 mol if the entropy change, ΔS, equals 5.42 J K^{-1}

$$\Delta S = nR \ln \frac{V_f}{V_i} \qquad (1.27)$$

3. Determine the molar Gibbs energy of reaction, $\Delta_r G$, at 40 °C when $\Delta_r H = -10.24$ kJ mol^{-1} and $\Delta_r S = +34.7$ J K^{-1} mol^{-1}

$$\Delta_r G = \Delta_r H - T \Delta_r S \qquad (1.28)$$

> Remember that you can't add two numbers together unless they have the same unit.

4. A typical carbonyl stretch occurs at 1780 cm^{-1}. Determine this value in Hz and kJ mol^{-1}.
5. The gas constant $R = 8.314462$ J K^{-1} mol^{-1} is also commonly reported with the unit dm^3 atm K^{-1} mol^{-1}. Determine the value with this unit.
6. The Beer–Lambert law states:

$$\log\frac{I_0}{I} = A = \varepsilon cl \tag{1.29}$$

5.00 mg of azulene ($C_{10}H_8$) is dissolved in 25 mL of cyclohexane, and 50 μL of this solution is added to 1 mL of cyclohexane. Calculate the expected absorption and proportion of transmitted light (I/I_0) if the molar extinction coefficient (ε) is 400 mol^{-1} dm^3 cm^{-1} at 580 nm in a 1 cm path length cuvette.

Remember that M is shorthand for mol dm^{-3}.

7. What is the pH of a 250 μM solution of uric acid? ($pK_a = 5.6$)

$$K_a = \frac{[H^+][A^-]}{[HA]} \tag{1.30}$$

Remember to convert all of your units to base SI.

8. The ideal gas equation, eqn (1.31), shows how the variables pressure, volume, number of moles and temperature relate to each other:

$$pV = nRT \tag{1.31}$$

Show that the units of R must be J K^{-1} mol^{-1}. Determine the pressure (in bar) exerted by 55.8 g of molecular nitrogen, N_2, in a box of 2.550 dm^3 at 20 °C.

9. The wavelength of emission is calibrated in a fluorimeter using the Raman (inelastic) scattering of light from a sample of water, which obeys the following equation:

$$\frac{1}{\lambda_{em}} = \frac{1}{\lambda_{ex}} - 0.340 \ \mu m^{-1} \tag{1.32}$$

Determine the wavelength of the scattered light, λ_{em} when incident light (λ_{ex}) of 350 nm is used.

Be careful with your units!

10. Determine the frequency of vibration, $\bar{v}$, in wavenumbers (cm^{-1}) of a $^{12}C^{16}O$ molecule, knowing that the bond strength $k = 1.902$ kN m^{-1} (c is the speed of light):

$$\bar{v} = \frac{1}{2\pi c}\sqrt{\frac{k}{\mu}} \tag{1.33}$$

The masses m_1 and m_2 are the masses of the individual atoms in the molecule.

$$\mu = \frac{m_1 m_2}{m_1 + m_2} \tag{1.34}$$

11. HCl fully dissociates in water. If 5.00 cm³ of a 38% w/w HCl solution (ρ = 1.189 kg dm⁻³) is made up to exactly 100 cm³ in water, what is the pH of the resulting solution? What mass of NaOH is required to neutralise 10.00 cm³ of this solution?

> w/w means weight weight, *i.e.* number of grams of solute in 100 g.

Solutions to Exercises

Solutions: Exercise 1.1

This is not really an 'answer' as such; more a demonstration that $n^0 = 1$

$$n^0 = n^{(3-3)} = n^3 \div n^3$$
$$= \frac{n^3}{n^3}$$
$$= \frac{n \times n \times n}{n \times n \times n}$$
$$= \frac{\cancel{n} \times \cancel{n} \times \cancel{n}}{\cancel{n} \times \cancel{n} \times \cancel{n}}$$
$$= \frac{1}{1} = 1$$

Solutions: Exercise 1.2

1. $\left(a^2\right)^3 = a^6$

2. $\left(b^7\right)^9 = b^{63}$

3. $c^3 \times c^8 = c^{11}$

4. $d^4 \times c^2 + d^4 \times f^3 = d^4\left(c^2 + f^3\right)$

   —further simplification not possible, owing to different bases

5. $g^{1/2} \times g^{3/2} = g^{4/2} = g^2$

6. $\dfrac{h^{1/3}}{h^{2/3}} = h^{-1/3}$

7. $j^{-2/7} \times j^{9/7} = j^{7/7} = j$

8. $\left(k^{4/5}\right)^5 = k^4$

9. $\left(m^{4/3}\right)^{-2} = m^{-8/3}$

Solutions: Exercise 1.3

1. $\log_3 a = 9; 3^9 = a = 19683$

2. $\log_5 125 = b; 5^b = 125, b = 3$

3. $\log_c 64 = 6; c^6 = 64, c = 2$

4. $\log_2 d + \log_2 f = 10; \log_2(df)$
   $= 10, 2^{10} = df = 1024$

5. $\log_g X = 4; g^4 = X$

6. $\log_6(h) = 4; 6^4 = h = 1296$

7. $a = 3^4; \log_3(a) = 4, a = 81$

8. $125 = b^3; \log_b(125) = 3, b = 5$

9. $32 = c^5; \log_c(32) = 5, c = 2$

10. $d = 7^3; \log_7(d) = 3, d = 343$

11. $625 = f^4; \log_f(625) = 4, f = 5$

12. $1024 = 2^g; \log_2(1024) = g = 10$

Solutions: Exercise 1.4

1. $T = \dfrac{pV}{nR}$

2. $T = \left(\dfrac{\Delta H - \Delta G}{\Delta S} \right)$

3. $\left[OH^- \right] = \dfrac{K\left[H_2O \right]}{\left[H^+ \right]}$

4. $b = \dfrac{(a-c)}{5}$

5. $f = \sqrt{\dfrac{(d+g)}{3}}$

6. $a = \dfrac{2(s-ut)}{t^2}$

7. $c = \dfrac{b^2}{a^2}$

8. $D = \sqrt{\dfrac{At}{\mu F}}$

9. $h = \dfrac{\left(S - 2\pi r^2 \right)}{2\pi r}$

10. $c = \dfrac{d(Q+1)}{(Q-1)}$

You almost certainly think of concentrations as mol dm^{-3}, but it is useful to remember the base SI unit as sometimes units need cancelling.

Solutions: Exercise 1.5

1. $F = $ kg m s^{-2} or N
2. $p = $ kg m^{-1} s^{-2} or Pa
3. $E = $ kg m^2 s^{-2} or J
4. $E = $ kg m^2 s^{-2} or J
5. $C_{p,m} = $ kg m^2 s^{-2} K^{-1} mol^{-1}
   or J K^{-1} mol^{-1}
6. $\left[A \right] = $ mol m^{-3}

7. $Q = $ A s or C
8. $V = $ kg m^2 s^{-3} A^{-1}
9. $R = $ kg m^2 s^{-3} A^{-2}
10. $\rho = $ kg m^3 s^{-3} A^{-2}
11. $\kappa = $ kg^{-1} m^{-3} s^3 A^2
12. $\Lambda_m = $ kg^{-1} s^3 A^2 mol^{-1}

Solutions: Exercise 1.6

1. 3.4×10^3 nm and 3.4×10^{-6} m
2. 22.4×10^{-3} m^3 and 22.4×10^3 cm^3
3. 0.2704 kg mol^{-1} and 0.4490 zg
4. 0.057 s^{-1} and 204 h^{-1}
5. 0.3400 µm^{-1} and 3.400×10^5 m^{-1}
6. 63.2 mol^{-1} dm^3 s^{-1}

"You don't need to include the molecule^{-1} as it is not a unit and is implicit" (Answer 3).

Solutions: Exercise 1.7

1. kg m s^{-1}
2. $V_f = 1.300$ V
3. $\Delta_r G = -21.10$ kJ mol^{-1}
4. 5.336×10^{13} Hz and 21.29 kJ mol^{-1}
5. $R = 0.0820574$ dm^3 atm K^{-1} mol^{-1}
6. $A = 0.0297$ and $\dfrac{I}{I_0} = 93.4\%$

7. pH 1.0
8. $p = 19.0$ bar
9. $\lambda_{em} = 397$ nm
10. $\tilde{\nu} = 2170$ cm^{-1}
11. pH = 0.207; $m_{NaOH} = 0.248$ g

Learning Points: What We'll Cover

- ☐ Recognising probabilities in chemistry
- ☐ Quantifying probabilities in the context of simple exercises
- ☐ Permutations (different rearrangements) of systems and how this affects observed results
- ☐ Relating distinguishable events in probability to observed states in chemistry
- ☐ Relating dependence of events in probability to models in chemistry
- ☐ Appearance of the standard distribution in probability studies
- ☐ Presenting data distributions as histograms and the effects of different presentation approaches
- ☐ Properties of the standard deviation
- ☐ Calculating a standard deviation and using it appropriately

Why This Chapter Is Important

- • The laws of probability and statistics govern all chemical processes; understanding these concepts can give insight into what is happening at a molecular level.
- • The observations we make are of the most probable outcome of a chemical process; this is a key factor in relating the microscopic world of atoms and molecules to the world in which we make measurements such as temperature.
- • Probability and statistics are central to making sense of experimental data.

Probability and Statistics in Chemistry

'Huh. What are the chances?' A phrase that is often heard and highlights the fact that the laws of probability are to be found everywhere, whether we are determining who plays first in a sporting competition to who wins a sudden windfall with a lottery ticket. Running into an old friend from school at a particular event? What are the chances? Having the same bus driver on the way home as you had that morning on the way to work? What are the chances?

In chemistry, we often look at outcomes of our reactions as definite outcomes, however they are all rooted in probability. The outcome we observe is simply the most probable outcome of all possibilities (as shown by the study of statistical mechanics and quantum chemistry), while macroscopic parameters, such as the Gibbs energy, arise from the second law of thermodynamics and the statistical nature of entropy.

Finally, when making measurements, we introduce a degree of error into our science; knowing how these uncertainties arise and how to deal with these random occurrences also makes use of the laws of probability to quantify errors in our measurements.

This chapter will focus on the probabilities behind chemical processes and how to use statistical outcomes to ascribe meaning to our experimental measurements.

Look into any chemical process and we find probability right at its heart. At the most basic level, a probability tells us the likelihood of something happening (Key Concept 2.1). Looking further, we see that there is a direct link between probability, statistics and thermodynamic entropy, driving processes such as crystallisation, dissolution, and diffusion and expansion of gases.

It can be challenging to see how probability fits into chemistry, and in particular how a probabilistic approach can predict reaction outcomes. However, you will have seen this probability in action already in chemical kinetics—the mathematics of first-order kinetics and even the Arrhenius equation itself have probability hard-coded into their identities.

---

❶ Key Concept 2.1 What is a Probability?

A probability indicates how likely something (often called an *event*) is to happen, and often carries the symbol *P*.[†]

Probability can vary between two values:

$$P = 0 \quad \text{Event will never happen.}$$
$$P = 1 \quad \text{Event will always happen.}$$

In reality, very few things are certain (*i.e.* $P = 1$), and conversely very few things are forbidden ($P = 0$). In our everyday scientific studies, we tend to consider events as probable (*i.e.* P approaches 1) or improbable (P approaches 0).

2.1 Probability in Chemical Kinetics

The Arrhenius equation is one of the most fundamental relationships in chemistry and we discuss it numerous times in this text; however, the depth of its meaning combined with its simplicity of expression makes it a fruitful example in many situations.

Let's first recall the general form of the Arrhenius equation:

$$k = Ae^{-E_a/RT} \tag{2.1}$$

This relates the rate constant, k, to the collision frequency, A, *via* a decaying exponential. We will devote our attention to the decaying exponential in this example. What are the limits of this decaying exponential; *i.e.* what is the biggest it can be, and what is the smallest it can be?

T is the thermodynamic temperature, *R* the gas constant and E_a the activation energy for the process.

For simplicity, let's rewrite our decaying exponential as:

$$e^{-E_a/RT} = \frac{1}{e^{E_a/RT}} \tag{2.2}$$

For a given reaction, E_a will be a constant and R is also a constant (the gas constant). Therefore, the only thing which affects the value of the decaying exponential is the temperature, T, which we impose on the system. By considering the highest and lowest values that T can have, we see that the exponential term in the Arrhenius equation will vary between zero and one $\left(e^{-E_a/R\times 0} = e^{-\infty} = 0\right.$

[†]Not to be confused with power which is often given the same symbol!

and $e^{-E_a/R \times \infty} = e^0 = 1$)—in other words, the fundamental relationship within the Arrhenius equation is a probability. As a reaction is cooled, it slows down until we remove all thermal energy, at which point there is zero chance of the reaction happening. Conversely, as we heat a reaction, it speeds up until there is near certainty of every collision resulting in a successful reaction. This approach illustrates one role of probabilities in chemistry; the mathematics behind it is shown in Toolkit A.7.

2.2 Flipping Coins, Filling Reactions

The most basic probability games that we are familiar with involve 'flipping coins'. "Heads or tails?" most exchanges begin, with all participants confidently assuming that the outcome is equally favoured either way.[‡] In chemistry, there are many situations which follow similar probabilities, and all have profound effects on the observed properties of a system.

Example 2.1 Outcomes of Coin Flipping

Q: What is the most common heads / tails outcome from flipping ten fair coins?

A: It will be no surprise that the most common outcome results from five heads and five tails. However, it is by no means a certainty that this will be the outcome; it is simply the *most common* outcome. Indeed, it is possible (though quite improbable!) that we get the outcome of ten heads!

It is perhaps easiest to visualise possible outcomes on a histogram:

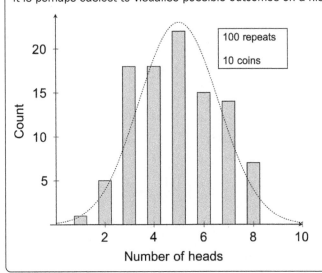

[‡]Of course, we are assuming that the coin is fair. Most of the time, this is a reasonable assumption; however, the work which goes into assuring a 'fair coin' is lengthy and outwith the scope of this text. The 'fairness' of a coin is of particular interest in the sporting community for deciding who goes first!

This histogram shows a series of randomised coin tosses; we tossed 10 coins and recorded the number of heads in the set. We then repeated the test 99 more times, each time recording the number of heads. Notice that the most common outcome was indeed with five heads. However, when we repeated the experiment, we observed the following:

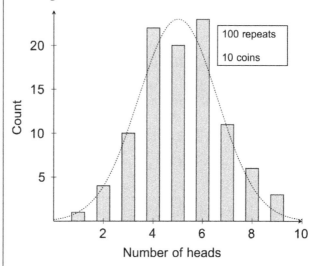

What we observe from this is that even though our instinct says what we should expect from our most probable result, it will not always turn out this way. However, when we look at the predicted probability distribution (dotted line), we see the same general pattern emerging.

As we increase the number of repeats, we get closer to this idealised probability distribution, as can be seen for a test with 10 000 repeats:

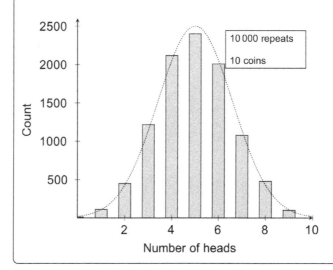

> ⊘ **Key Concept 2.2 The Mathematics of Coin Flipping**
>
> The probabilities of flipping a coin are similar to many other processes, and they may be determined as follows:
>
> - The total number of combinations (permutations) from flipping N coins may be found by evaluating 2^N.
> - The number of permutations for a given observed outcome of N coins (the multiplicity, Ω) may be found by evaluating $\Omega = N!/(n_H! \times n_T!)$.[§]
>
> These quantities may be readily adapted for other situations with more than two outcomes; for outcomes of six-sided dice (six outcomes per die), the equivalent expressions become 6^N and $\Omega = N!/(n_1! \times n_2! \times \ldots \times n_6!)$.

Permutations consider the order in which the result is achieved, so THH is a different permutation from HTH.

An observed outcome is 'two heads, one tail', where HHT, HTH, etc., are different permutations giving the same observed outcome.

> ⊘ **Key Concept 2.3 Relating Number of Outcomes to Probability**
>
> The probability of a given outcome can be determined by knowing the number of ways of achieving that outcome (the multiplicity, Ω) and the total number of possible permutations (rearrangements) of the system.
>
> $$\text{Probability of given outcome, } P = \frac{\text{Multiplicity of the outcome, } \Omega}{\text{Total number of permutations, } N}$$
>
> To determine the probability of achieving 'three heads, one tail' from four coins, we know that there are $\Omega = 4$ ways to achieve this result (Table 2.1), and the total number of permutations $N = 2^4$ (Key Concept 2.5); the probability of achieving this result is:
>
> $$P_{3H1T} = \frac{4}{16} = \frac{1}{4}$$

The symbol ! is called a factorial; you can find out more about this mathematical function in Toolkit A.11.

❓**Exercise 2.1 Calculation of Probabilities**

For each example, evaluate the multiplicity (the number of permutations for the observed state) and determine the probability of this state occurring by calculating the overall number of possible permutations. You will need to use the relations in Key Concept 2.2 and you may need to adapt these for the different circumstances.

1. Five coins flipped, resulting in an observed state of three heads, two tails.

[§]H = head, T = tail and N is the total number of outcomes, whereas n_H and n_T are the total number of heads and tails, respectively.

2. One coin is flipped four times, giving the result HTTH in that order.
3. The most common outcome when tossing eight coins.
4. Four six-sided dice (D6) are rolled, resulting in an observed state of two '3's, one '4' and one '5'.
5. Four four-sided dice (D4) are rolled, resulting in an observed state of one each of '1', '2', '3' and '4'.

Coin flips are the classic way of demonstrating probability exercises, with most readers being familiar with this exercise. The mathematics behind the probability of this process is shown in Key Concept 2.2. We can flip coins all day long (and, as authors, we have spent more time flipping coins than is strictly good for us!); however, this does not necessarily help us visualise the chemistry of the situation.

Let us instead picture a situation in which we are filling a reaction chamber with an ideal gas very slowly. Our reaction chamber is of a rather special design; it consists of two joined chambers, A and B (Figure 2.1). We are filling it so slowly that we are only introducing a single molecule at a time. How then do the gas molecules spread themselves around?

With a few basic assumptions (free movement of molecules through system, molecules are only found in chambers and not in the connecting tube and there are no interactions between molecules), it can be seen that each molecule will have an equal probability of occupying either of the joined chambers. The first molecule has a probability of 0.5 of entering chamber A and a probability of 0.5 of entering chamber B. The same will be true of the second, third and every subsequent molecule. Since we are working with an ideal gas, there are no interactions between gas molecules, so the presence of a molecule in one chamber does not affect the ability of another gas molecule to enter.

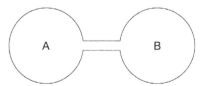

Figure 2.1 A twin-chambered reaction vessel. For the purposes of this example, we assume that molecules can only occupy either chamber A or chamber B and can pass freely between the chambers at any point, but cannot exist in the connecting tube.

With respect to the probabilities, this is the same problem as the coin-flipping example already explored; each molecule can exist in one of two states; A or B (remember that each coin can exist in one of two states, H or T). Consequently we should expect the most probable state to be one in which both chambers have an equal number of molecules inside. While instinct tells us that we should expect the gas molecules to be evenly distributed, the reality is that because there is an inherently probabilistic process, the distribution will be *approximately* equal, but not necessarily *exactly* equal.

If we allow four gas molecules to enter our reaction chamber under the free movement principles, there are five observable states; either we have all in chamber A, all in chamber B, or some variant distribution between the two (all possible outcomes are listed in Table 2.1). However, not all five states are equally probable.

Since this example has the same mathematics as the coin flip, adding each additional molecule carries a probability $P = 0.5$ of entering chamber A or chamber B. To visualise how each of the states of this four-molecule system arise, we can construct a probability tree, as shown in Figure 2.2.

Immediately, we can see that there are 16 possible paths (2^4), but only five possible observable results. The *permutations* for each of these observables are shown in Table 2.1 and we see that the outcome with the molecules distributed equally between the two chambers has the most possible permutations, so is the most probable single outcome.

Table 2.1 The different permutations which give rise to each of the possible observable states of the system shown in Figure 2.1.

A_4,B_0	A_3,B_1	A_2,B_2	A_1,B_3	A_0,B_4
AAAA	AAAB	AABB	ABBB	BBBB
	AABA	ABAB	BABB	
	ABAA	ABBA	BBAB	
	BAAA	BAAB	BBBA	
		BABA		
		BBAA		
$P = \frac{1}{16}$	$P = \frac{1}{4}$	$P = \frac{3}{8}$	$P = \frac{1}{4}$	$P = \frac{1}{16}$

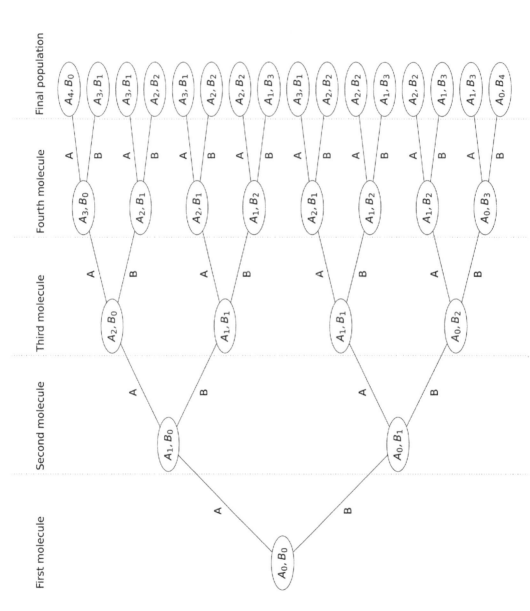

Figure 2.2 Each molecule has equal probability of going into chamber A or chamber B; consequently, we can evaluate the probabilities of each of the outcomes as we add each molecule.

This principle of *combinatorics* is central to the field of *statistical mechanics,* connecting the behaviour of individual molecules to the observable measurements which we can make. Fundamentally, the outcomes of our chemical reactions are governed by probability, with the most likely outcome (the most likely *macro*state) being that with the most possible rearrangements (the most possible *micro*states). The connection between observed states and distinguishable events is detailed in Key Concept 2.4.

---

❶ Key Concept 2.4 Distinguishable Events and Observed States

Events can be either *distinguishable* or *indistinguishable*. In the context of a four-coin toss:

- A *distinguishable* event describes the overall outcome, regardless of how it is achieved; for example, 'three heads, one tail' is easily distinguishable from 'two heads, two tails'.
- An *indistinguishable* event cannot be distinguished by the observer; for example, the outcome 'three heads, one tail' can be formed in four different ways (HHHT, HHTH, HTHH, THHH), none of which can be distinguished without labelling the coins.

In the context of chemistry, we use the term *states* as follows:

- A *macrostate* is a *distinguishable outcome*, describing the number of molecules in each given state.
- A *microstate* is an *indistinguishable outcome*, detailing the number of ways of creating an observed *macrostate*.

The ratio of microstates to macrostates is central to calculations of statistical entropy.

It is easy to get confused between the prefixes *micro* and *macro*. Remember that *macro* means 'big' or 'something we can see', while *micro* means 'small' or 'something we can't see'. This should help to keep the terms separate!

Another example of how system rearrangements affect observed outcomes can be seen when we consider the third law of thermodynamics: 'The entropy of a perfect crystal at 0 K is zero'. This is connected to the logarithmic nature of statistical entropy, which we discussed in Section 1.2.6 and represented in eqn (1.17). A perfect crystal has all molecules perfectly arranged, with no possible way of reordering the positions of these molecules. This means that its multiplicity $\Omega = 1$ and its statistical entropy is zero ($\ln \Omega = \ln 1 = 0$).

For a molecule such as carbon monoxide, there is an element of randomness in the crystal when it is cooled and crystallises, owing to the similarity in size and charge distribution across the carbon

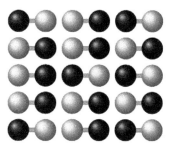

Figure 2.3 A carbon monoxide crystal can show residual entropy, owing to the small dipole moment and the negligible difference in energy between a (CO CO) arrangement and a (CO OC) arrangement. This 2-dimensional example shows 15 molecules in one of the $2^{15} = 32\,768$ different permutations available to this 2D system.

and oxygen atoms as illustrated in Figure 2.3. This gives rise to a phenomenon known as *residual entropy*, which can be calculated as shown in Example 2.2.

💬
Here, we are using the 'rules of logs', covered in Key Concept 1.5

💬
An alternate form of the Boltzmann formula is $S = k_B \ln W$, where W is used instead of Ω. W is functionally the same as Ω, but with a basis in energetics rather than statistics.

Example 2.2 Residual Entropy in a Carbon Monoxide Crystal

The 2-dimensional CO crystalline lattice shown in Figure 2.3 illustrates the similarity of size between the carbon and oxygen atoms in carbon monoxide, together with the possibility of random orientation of each molecule within the rigid crystal lattice.

As each CO molecule in a crystal can adopt one of two orientations (CO or OC), the mathematics of this system is the same as the coin-flipping example; each molecule contributes two possible outcomes, each independent of any other molecule in the lattice. For a lattice of N molecules, each with p possible equal orientations, the number of possible rearrangements (the multiplicity) Ω can be found as:

$$\text{Number of possible rearrangements, } \Omega = p^N$$

with the entropy being found as:

$$S = k_B \ln \Omega = k_B \ln (p^N) = N k_B \ln p$$

For the 15-molecule 2D lattice shown in Figure 2.3 there are 2^{15} possible states, giving a residual statistical entropy of $S = k_B \ln (2^{15}) = 15 k_B \ln 2$.

❓Exercise 2.2 Calculation of Chemistry Probability

For each of the situations described, calculate the multiplicity and the statistical entropy as a scale of the Boltzmann constant, k_B. You will need to use the mathematics covered in Key Concept 2.2.

1. Two interlinked gas vessels populated with 12 molecules, distributed equally between the chambers.
2. Three interlinked gas vessels populated with 12 molecules, distributed equally between the chambers.
3. Four interlinked gas vessels populated with 12 molecules, distributed equally between the chambers.
4. Four interlinked gas vessels populated with 10 molecules; one in the first chamber, two in the second, three in the third and four in the fourth.

2.2.1 Dependence of Events

In the examples seen so far, we have only looked at events which are *independent*. This is when the outcome of each event (or its probability) does not depend at all on the outcome of any previous event. The example of the coin toss or 'reaction vessel filling' in Section 2.2 is an example of this. *Dependent* events, however, are affected by the outcome of any previous event; the probability of which ball is drawn at which point in a lottery draw depends on which balls have been drawn before; for example, the probability of the '42' ball being drawn first is different from the probability of it being drawn fourth, as there are three fewer balls in the machine at the time of the fourth ball being drawn. The same principle applies for drawing from a shuffled deck of cards.

An example of a chemical setting for *dependent* events is the packing of a crystal lattice. Each molecule can only occupy one lattice site, and there is a finite known number of lattice sites. The statistical entropy of the system is related to the number of possible rearrangements of this lattice; being able to calculate the number of possible *permutations* is vital to our statistical understanding, as this allows us to determine the probability of a given event (Key Concept 2.3).

To appreciate the statistical nature of our chemical systems, we need to be able to calculate the number of *permutations* for our observed states, as the probability of something happening is intrinsically related to the number of possible permutations. We show how to calculate the number of permutations of dependent and independent events in Key Concepts 2.5 and 2.6.

❶ Key Concept 2.5 Permutations of Independent Events

For a series of a independent events, where each event has p possible outcomes, the total number of possible permutations (rearrangements), N, may be found by the formula:

$$\text{Total number of permutations, } N = p^a$$

This is the general case for the example shown in Key Concept 2.2. In the context of rolling a six-sided die four times, each event has $p = 6$ possible outcomes, and we are observing a series of $a = 4$ independent events. Therefore, $N = 6^4 = 1296$ possible permutations.

The exclaimation point after a number, $p!$ indicates a *factorial* number; that is the number multiplied by every number lower than it. *e.g.* $4! = 4 \times 3 \times 2 \times 1$. See Toolkit A.11 for a more complete explanation.

❶ Key Concept 2.6 Permutations of Dependent Events

For a series of n dependent events, where the first event has p possible outcomes and p is reduced by each subsequent event, the total number of possible permutations, N may be found by the formula:

$$\text{Total number of outcomes, } N = \frac{p!}{n! \times (p-n)!}$$

In the context of drawing a specific combination of $n = 3$ cards from a deck of $p = 10$ cards, the number of possible combinations can be found *via*:

$$\text{Total number of outcomes, } N = \frac{10!}{3!7!} = 120$$

The probability of any specific outcome becomes the reciprocal of this value:

$$\text{Probability of a single outcome, } P = \frac{1}{\text{Total number of outcomes, } N}$$

Knowing how to calculate these allows us to determine statistical properties of chemical systems, whether we are quantifying microstates for a given arrangement of molecules or determining the statistical entropy of a system to predict the direction of change in accordance with the laws of thermodynamics.

❓Exercise 2.3 Calculation of Dependent Events

1. A lottery draw is configured so that five numbers are chosen from 40 possible values. Calculate the number of possible outcomes and hence the probability of winning the jackpot.

2. A gas can be modelled as a lattice with n sites; a site may be either occupied by a single molecule or unoccupied. Calculate the number of possible permutations for a lattice of six sites populated with four gas molecules.

2.2.2 Calculations of Chemical Systems: Stirling's Approximation

When we are determining the number of rearrangements of coins or cards, we are dealing with relatively small systems *i.e.* no more than a handful of coins or a deck of cards. When we wish to consider statistical factors in chemical systems, we are dealing with considerably larger values. Even a quantum dot—arguably one of the smallest solid state systems which we could consider as 'chemical'—we are dealing with hundreds of atoms. By the time we are considering anything approaching an observable quantity of material, we will have upwards of 10^{18} atoms to consider. When we come to determine a factorial, *i.e.* $10^{18}!$, we will rapidly find that our computers (let alone our calculators!) run out of memory space to handle values so large. To handle such large factorial calculations, we use an approximation known as 'Stirling's approximation' (Key Concept 2.7).

---

❶ Key Concept 2.7 Stirling's Approximation

Stirling's approximation allows for the manipulation of factorials of extremely large factorial calculations. The logarithmic form of the approximation arises from the derivation itself, but this is readily rearranged to obtain a factorial in direct form.

Its derivation is available in *The Chemistry Maths Book* by Erich Steiner.

Logarithmic form:

$$\ln(x!) \approx x \ln(x) - x$$

Direct form:

$$x! \approx \left(\frac{x}{e}\right)^x$$

Example 2.3 Evaluating Probabilities of Molecular Systems

We revisit the linked gas vessels from earlier in the chapter but now our calculations will be on a realistic molecular scale.

Q: Four interlinked gas vessels are set up as before, where gas molecules are free to pass between the vessels; 6×10^{20} (approximately 1 mmol) atoms of helium are introduced into this system. Calculate the statistical entropy of the state in which the molecules are distributed equally between the chambers.

A: In the earlier example, we saw that the overall number of permutations could be found *via* the formula p^a; in this case, each event (the introduction of each He atom) has four possible outcomes (it can enter one of four chambers), so $p = 4$, and this is repeated for each molecule, so $a = 6 \times 10^{20}$.

We wish to calculate the statistical entropy *via* $S = k_B \ln \Omega$, so the value we need to determine is Ω. In Key Concept 2.2, we showed how the multiplicity of a given state may be calculated as:

$$\Omega = \frac{N!}{n_1! n_2! n_3! \ldots}$$

In this example, a He atom can occupy any of the four gas vessels, so we identify how many atoms are in each one. We have said that we have an equal distribution, so each vessel has 1.5×10^{20} He atoms. This gives us a multiplicity calculation involving very large factorials!

$$\Omega_{He} = \frac{\left(6 \times 10^{20}\right)!}{\left(1.5 \times 10^{20}\right)! \times \left(1.5 \times 10^{20}\right)! \times \left(1.5 \times 10^{20}\right)! \times \left(1.5 \times 10^{20}\right)!}$$

For the purposes of applying Stirling's approximation, it will be helpful to use a substitution; we will say that there are n helium atoms in each chamber, and $4n$ atoms overall (so $n = 1.5 \times 10^{20}$ atoms). This greatly simplifies our expression for the multiplicity:

$$\Omega_{He} = \frac{(4n)!}{n!n!n!n!}$$

We recall that Stirling's approximation may be expressed in its direct form as:

$$x! \approx \left(\frac{x}{e}\right)^x$$

This allows us to write the multiplicity in a form of the approximation:

$$\Omega_{He} = \frac{\left(\dfrac{4n}{e}\right)^{4n}}{\left(\dfrac{n}{e}\right)^n \times \left(\dfrac{n}{e}\right)^n \times \left(\dfrac{n}{e}\right)^n \times \left(\dfrac{n}{e}\right)^n}$$

$$= \frac{\left(\dfrac{4n}{e}\right)^{4n}}{\left[\left(\dfrac{n}{e}\right)^n\right]^4}$$

We now group our powers together and we can then do some cancelling:

$$\Omega_{He} = \frac{\left(\dfrac{4n}{e}\right)^{4n}}{\left[\left(\dfrac{n}{e}\right)^{n}\right]^{4}}$$

$$= \frac{\left(\dfrac{4n}{e}\right)^{4n}}{\left(\dfrac{n}{e}\right)^{4n}}$$

$$= \left(\frac{\dfrac{4n}{e}}{\dfrac{n}{e}}\right)^{4n}$$

$$= \left(\frac{4n}{e} \times \frac{e}{n}\right)^{4n} = \left(\frac{4\cancel{n}}{\cancel{e}} \times \frac{\cancel{e}}{\cancel{n}}\right)^{4n}$$

$$= \left(4\right)^{4n}$$

This now looks almost simple! When we apply the Boltzmann relation to find the statistical entropy, we are taking a natural logarithm of this value:

$$S = k_B \ln \Omega$$

$$= k_B \ln \left(4\right)^{4n}$$

$$= k_B \times 4n \ln 4$$

Now we substitute our value of x back into the equation to determine our statistical entropy:

$$S = k_B \times 4n \ln 4$$

$$\approx 8.3 \times 10^{20} k_B$$

$$\approx 1.1 \times 10^{-2} \, \text{J K}^{-1}$$

2.3 Standard Distributions and Standard Deviations

In Example 2.1, we introduced the idea that when we do an experiment we expect to have a *distribution* of results. If we are making a series of observations on a fair coin (two possible outcomes), we would expect the most common single result (the 'peak') to be that of equal numbers of heads and tails. However, we saw that this is not a guarantee, and sometimes the most common result obtained may not be the same as that predicted. The same is true for our chemical systems; when gas molecules occupy a two-chambered vessel, we would expect the most common result to arise with half of the gas molecules in one chamber and half in the other, simply because there are more ways of achieving this result. With more trials, the more our data gather around a mean value, and the same is true of our experimental data. In any practical science, it is good practice to measure a value more than once (Chapter 3); either we measure the same system a number of times, or we make changes to the system and plot a graph of the results (Chapter 4).

If we consider a class of students asked to determine the conductivity of a 2% solution of sodium chloride (0.342 mol dm^{-3}), we would not expect every student to achieve the same value because of the random fluctuations inherent in experimental science. However, we would expect the class 'average' to be close to the accepted literature value of 30.2 mS cm^{-1},[¶] and we can infer meaning in our experiments from the way in which the data are distributed around this mean value. First of all, we need to discuss how data are *distributed* and what we mean by concepts such as 'average' (Key Concept 2.8).

⊕ Key Concept 2.8 Calculating an Average

In statistics, three averages must be considered.

1. The *mean*. This is frequently called the 'average' by most; it is the sum of all values divided by the number of values, N. Mathematically:

$$\mu_x = \frac{1}{N} \sum_{i=1}^{N} x_i$$

Note that μ represents a population mean, while $\bar{x}$ is used to represent a sample mean.

2. The *median*. This is the value 'in the middle'; to find this, sort all values from smallest to largest and identify the value in the 'middle'. In a sample of seven values (an odd number), rank the values in ascending order, and the fourth value in the list (the value in the middle of the list) will be the median value.

   If there are eight values (an even number), the median value will be the *mean* of the fourth and fifth values (the two values in the middle of the list).

3. The *mode*. This is the value which appears most frequently in a data series.

In the standard distribution, the mean, median and mode are all coincident at the peak of the distribution.

[¶] Electrical Conductivity of Aqueous Solutions, in *CRC Handbook of Chemistry and Physics*, CRC, Boca Raton, 100th edn.

?Exercise 2.4 Calculation of Averages

Determine the mean, median and the mode for each of the following data sets.

1. NaCl conductivity/mS cm^{-1}: 70.4, 69.8, 69.5, 70.3, 70.1, 69.5, 70.2, 70.4, 69.6, 70.1, 69.5, 69.6, 69.7, 69.8, 70.5.
2. (–)-menthone boiling point/°C: 204, 211, 212, 214, 205, 198, 204, 208, 207, 202, 203, 210, 205, 205, 202.
3. $\Delta_c H°$ ethanol/kJ mol^{-1}: –1216, –1211, –1219, –1315, –1277, –1307, –1333, –1312, –1270, –1329, –1387, –1273, –1211, –1293.

2.3.1 Distribution of Data

To view how data are distributed, we need to examine our data and quantify how many times each value appears. This is termed the *frequency* of the value. A graph of a distribution is simply a plot of how many times each value appears; the value on the x-axis and its frequency on the y-axis.

Depending on the precision of our measurements and the range over which they were made, the *frequency* of each data point may be higher or lower. Very precise measurements may result in each value only appearing once, and our 'distribution' will have very little meaning. For this reason, we use a technique called 'binning' to group values together, and we simply count how many values fall into each 'bin'; this frequency is then used to generate a histogram of the data, showing its distribution.

> In this case, precision is a measure of how many significant figures we can record.

Consider the data set listed in Table 2.2. These data are all unique values, so plotting a histogram will simply show a flat distribution. However, if we 'bin' them together at different 'bin' sizes, we see that the population distribution changes; with three bins, we still have a relatively flat distribution, but we see that we have a definite peak with four bins.

When we 'bin' our data, we split the range over which our data are spread into equal-sized 'chunks' and then we count how many data points fall into each 'bin'. In Table 2.2, we have a range of (15 – 12.3) = 2.7; if we divide this into three bins, each bin covers a range of 0.9, while dividing into four bins lets each bin cover a slightly smaller range, of 0.675. We can see the effect of changing the number of bins in Figure 2.4.

> Think of 'bins' as 'buckets', collecting values together; the bigger the bin, the more values it will collect together, but the fewer bins we will have. This has the effect of increasing frequencies for a group in order to see a meaningful distribution, but reduces the 'resolution' of the plot.

Let's go back to the conductivity data collected by our class of students and look at what happens when we start to combine the data from every student into histograms. When the data from a single student (two data points) are plotted on a histogram, we do not see a distribution. When we start to use larger data sets, however, we

Bins are always a constant size, so they are counting the number of values in a given range. Typically bin size is equal to (top value − bottom value)/ (number of bins).

It is tempting to use fewer bins to see the distribution more easily with a smaller data set; however, we are less able to visualise the 'spread' of the data set.

Table 2.2 For the same data set, the distribution between bins depends on the 'bin size'. With four bins, we have a smaller bin size, but a very different distribution of the data points.

Value	3 bins Bin size = 0.9 Frequency	4 bins Bin size = 0.675 Frequency
12.3	3 values	2 values
12.7		
13.2		1 value
13.8	3 values	4 values
14		
14.1		
14.2		
14.6	4 values	3 values
14.7		
15		

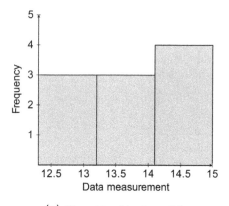

(a) Three bins, bin size = 0.9

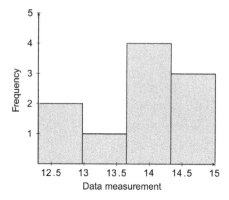

(b) Four bins, bin size = 0.675

Figure 2.4 In plotting the histograms from the data shown in Table 2.2, we see that, in this case, by increasing the number of bins (and slightly reducing the bin size as a result), we obtain a distribution with a clear peak.

start to see the distribution in our histograms. In Figure 2.5, we can see that the number of bins we use has an effect on how well we can visualise the distribution.

Figure 2.5 shows that with a large data set we start to *approximate* a symmetric distribution, but it will never exactly match. The more values we record and the smaller our bin sizes, the closer we will come to matching the classic 'bell curve' shape of the 'standard distribution'.

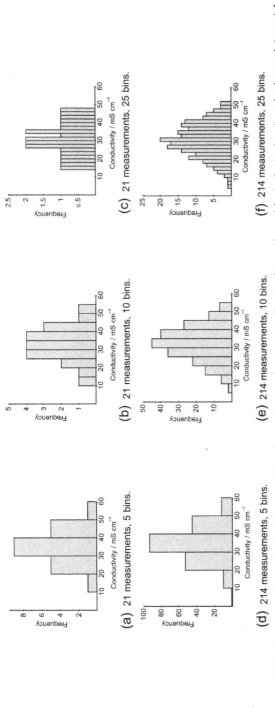

Figure 2.5 Comparing the effect of different bin sizes when we have a small set of data from 10 students (a–c) and a large data set from a class of 100 students (d–f). In general, we can see an approximation to a 'bell curve' with a few large bins for both, but we only see the extreme outliers in the data sets when we use lots of smaller bins.

> The 'standard' distribution is also frequently known as a 'normal' distribution or a 'Gaussian' distribution. There are mathematical differences between these; however, they all share the same mathematical definition $d(x) = Ae^{-kx^2}$ and, as such, the terms 'Gaussian', 'standard' and 'normal' are largely interchanged to mean the same thing when referring to distributions, though a pure mathematician may have strong opinions on this matter!

2.3.2 The 'Standard' Distribution

Any single measurement which is repeated often enough will eventually give a pattern approximating a standard distribution; most will have heard of this as a 'bell curve' because of its shape. This can occur for any measurement, whether we are measuring the conductivity of an aqueous solution of sodium chloride, counting how many 'heads' we obtain in a sequence of coin flips or even recording the height of every student attending a chemistry lecture. When we record a histogram of our data, the shape of the histogram becomes a better approximation to this 'perfect' standard distribution the more measurements we make.

The 'standard' distribution has two main characteristics, which make it useful for our statistical analysis. Firstly, results are clustered symmetrically around the mean value (denoted with a 'bar'; if our variable is x found at the peak of the distribution, the mean is denoted $\bar{x}$, 'x-bar'), and secondly the 'spread' of the data can be defined by a single parameter called the *standard deviation*, denoted using the symbol σ (the Greek letter 'sigma'). These two characteristics are shown in Figure 2.6, where we see how the standard deviation acts to enclose a fixed percentage of the population.

> There are a number of named distributions, including a *Poisson distribution*, very few of which are symmetric. The Maxwell–Boltzmann distribution of molecular energies is one example of an asymmetric distribution, and in these asymmetric distributions, the mean, mode and median values *do not* overlap at the peak.

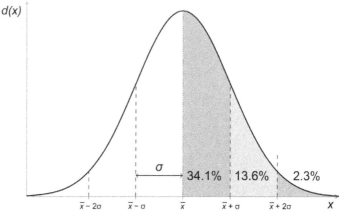

Figure 2.6   The classic 'bell' shape of the normal distribution. The distribution is symmetrical about the mean value, $\bar{x}$, while the standard deviation, σ, is a series of steps away from the mean value, each of which enclose a fixed percentage of the measurements: 34.1% of the results will be within one standard deviation above the mean, (34.1 + 13.6) = 47.7% will be within two standard deviations above the mean and so on, while the same proportions apply to standard deviations below the mean.

Put simply, if we can capture enough data, no matter what single parameter we are measuring, we should expect to obtain a histogram which should show a better fit to the standard distribution the more data points we measure.

2.3.3   The Standard Deviation

We mentioned the term 'standard deviation' in Section 2.3.2 and stated that it was a characteristic of the standard distribution such that it enclosed a fixed percentage of the data series. Figure 2.6 shows the percentages which fit within the first two standard deviations above the mean (and, by symmetry, these are the same values for the first two standard deviations below the mean); however, the standard deviation goes even further than this, quantifying what proportion of the data series lies within higher multiples of the standard deviation. A list of the proportions of a population enclosed by each successive standard deviation is shown in Table 2.3. This property of the standard deviation makes it exceptionally powerful in analysing our data and quantifying errors. We talk more about this application in Chapter 3.

Calculating the standard deviation is fairly straight forward (if laborious to do by hand); its formula is:

$$\sigma = \sqrt{\frac{\sum_{i=1}^{N}\left(x_i - \mu\right)^2}{N}} \tag{2.3}$$

This standard deviation is often also called the *population* standard deviation, as it is intended to cover all data possible, and uses the *population mean, μ*. In our experiments, we record considerably less data than this (even if we record 30 values, we are still only working with a *sample* of the population), so we use a different expression, the *sample* standard deviation:

Table 2.3   Each standard deviation from the mean encloses a fixed proportion of the measurements; thus over 95% of the data lie within two standard deviations of the mean (47.725% above the mean and 47.725% below the mean).

Interval	Proportion within interval / %
1σ	68.269
2σ	95.450
3σ	99.730
4σ	99.994
5σ	99.999 943
6σ	99.999 999 802

$$s = \sqrt{\frac{\sum_{i=1}^{N} \left(x_i - \bar{x} \right)^2}{N-1}}$$

(2.4)

The sample standard deviation differs because it is recognising that, by just taking a sample, we do not know the exact value of the population mean, or 'the real value'. As the sample size increases, our confidence in the value of the mean increases and the values of the standard deviation and the sample standard deviation converge (as it makes less difference whether we are dividing by N or $N-1$).

These equations seem to be complex and potentially impenetrable, so let's break them down into plain English to follow through the mathematical instructions. We will focus on eqn (2.3) first, and then explain how eqn (2.4) differs.

1. We have a population of N measurements. We find the *mean* (the 'average') of these values; this gives us the population average μ.
2. We subtract μ from each measurement; for the first value x_1 we subtract μ, then we subtract μ from the second value (x_2), again for the third, and so on, all the way to the last value, the Nth value, x_N. The difference between a value and the mean is called the *residual*.
3. We then square all of these *residuals* (yielding only positive numbers) and add them all up (the summation).
4. Finally, we divide by the number of measurements in the population, N, and find the square root of the result.

> ⬛
>
> A summation works by saying, 'Evaluate the expression for each value of x and then add all the expressions up.' The summation
>
> $$\sum_{i=1}^{N} x_i^2 + 1 \text{ can}$$
>
> 'translate' as 'take the first value (x_1), square it and add one, repeat this for the next (x_2) and each value up to the last one, then add all the results together.'

The *sample* standard deviation shown in eqn (2.4) differs only in that we now use a *sample mean*, $\bar{x}$, to determine the residual, and we divide by ($N - 1$); this has the effect of slightly increasing the value of the deviation, reflecting the more limited nature of a data sample compared with a population.

The standard deviation is an essential tool in statistics and is central in analysis of experimental uncertainty; we cover this further in Chapter 3.

❓Exercise 2.5 Calculation of Standard Deviations

For each of the data series listed, calculate the sample mean and the sample standard deviation.

1. $\Delta_{dil}H$ HCl/kJ mol^{-1}: −1.966, −1.732, −1.546, −1.423, −1.454, −1.842, −1.224, −2.095, −1.858, −1.769.
2. Surface tension of H_2O (25 °C)/mN m^{-1}: 76.54, 82.90, 72.64, 71.27, 70.22, 73.21, 75.68, 79.91, 78.01, 73.47.

2.3.4 Using the Standard Deviation

We have seen from our exploration of the standard distribution that we need a large number of values and that we must group them appropriately in order to see anything approximating the familiar shape of the standard distribution. For this reason, calculating a standard deviation on any fewer than ten or so values becomes rather dubious; fitting a standard distribution to five values (such as might be recorded in a short laboratory exercise) does not make sense. Consider the students' data shown in Table 2.4; this is a series of measurements made by a single student of the equilibrium constant for the association of iron(III) with the thiocyanate anion

$$\text{Fe}^{3+} + \text{SCN}^- \rightleftharpoons \text{Fe(SCN)}^{2+} \tag{2.5}$$

We could treat these data just like any other statistical data, calculating the sample mean ($\bar{x}$) and the sample standard deviation (s); however, it is worth appraising these in the context of the data distribution. We can represent these data as a histogram; we will group into five 'bins' and count how many values fall into each bin. Because there are so few data values, we have very low frequencies and it does not make much sense to have many more bins.

When we take the histogram data and use the calculated mean and standard deviation to model a standard distribution, a comparison of the two (Figure 2.7) shows that, while there is a (very) loose correlation, there is no guarantee that any further data collected will continue to fit the same distribution curve (indeed, it is almost certain not to!). This means that any determination of a mean or standard deviation on such a small sample set will be inaccurate for a much larger data set. Instead, the standard deviation of a data set is used to determine a standard error; we discuss this further in Chapter 3.

Table 2.4 Five measurements of the equilibrium constant for iron(III) thiocyanate (students' data from Swansea University). Statistical parameters $\bar{x}$ (sample mean) and s (sample standard deviation) are shown.

Measurement	K_{eq} / mol^{-1} dm^3
1	318
2	267
3	265
4	264
5	281

$\bar{x}$ = 279 mol^{-1} dm^3; s = 23 mol^{-1} dm^3

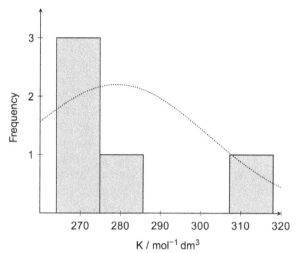

Figure 2.7   A histogram of the data shown in Table 2.4, together with a model of the standard distribution (dashed line) based on the mean and standard deviation of these values. Five bins selected, bin size = 10.8 mol dm^{-3}.

2.4 Summary

Probability is central to chemical processes, whether we consider how molecules interact in chemical reactions or the likelihood of observing particular states. The principles of probability can be used to derive models of chemical systems to fit observations and is central to the field of statistical mechanics; connecting the quantum world of wave functions and probabilities with the observable world of thermodynamics and kinetics.

Statistics are also a vital component of error analysis; however, it is useful to understand the limitations of statistical analysis, so that we may appraise our results appropriately. We will talk more about this in Chapter 3.

Summary: Chapter 2

Probabilities

- For a independent events (e.g. coin tosses, die rolls), where each event has p outcomes:

$$\text{Number of permutations, } N_p = p^a$$

- For a given observed outcome from N independent events (e.g. two heads, three tails) with p outcomes:

$$\text{Multiplicity, } \Omega = \frac{N!}{n_1! \times n_2! \times \ldots \times n_p!}$$

- The probability, P, of an observed outcome of independent events may be found from its multiplicity Ω and the total number of permutations, N_p, by

$$P = \frac{\Omega}{N_p}$$

- Macrostates are examples of distinguishable outcomes in molecular systems.
- Microstates are examples of indistinguishable outcomes in molecular systems.
- Stirling's approximation allows for statistical consideration of large systems.

Statistics

- The term 'average' is *normally* taken to be the mean, but the 'mode' and 'median' are also averages.
- Visualisations of data distributions approximate a standard distribution with increasing numbers of measurements.
- The standard deviation is a description of the 'spread' of data, and defines fixed proportions of the data set from the mean. One standard deviation from the mean will encompass 68.3% of the data, two standard deviations from the mean will encompass 95.5% of the data, and so on.
- For small data sets, where the 'mean value' is calculated from the same data *(sample mean, $\bar{x}$)*, the *sample standard deviation, s,* should be used

$$s = \sqrt{\frac{\sum_{i=1}^{N}(x_i - \bar{x})^2}{N-1}}$$

- For large data sets in which the 'mean' is known externally (the 'true' value from the literature or another appropriate source, denoted μ), the *population standard deviation, σ,* should be used

$$\sigma = \sqrt{\frac{\sum_{i=1}^{N}(x_i - \mu)^2}{N}}$$

- A sample standard deviation on a small data set (<10 points) can give spurious statistical results, but can give an indication of the uncertainty in an experiment (see Chapter 3.)

? Exercise 2.6 Probability and Statistics in a Chemical Context

1. A mass of 2.91×10^{-21} g SF_6 is introduced into a container of volume 7.17×10^{-19} cm^3 at standard ambient temperature and pressure. By determining the molecular volume of SF_6 from the molar volume of a gas under these conditions, determine the number of lattice sites available to the molecules and, from the number of molecules present, determine the multiplicity of the system.

2. A four-chambered vessel is manufactured to handle small volumes of gas; the chambers are connected in such a way that molecules are able to pass freely from one chamber to another. Twelve gas molecules are introduced to this system. Assuming that gas molecules are only present in the four chambers and not in the interconnecting tubes, determine the probability of having exactly three gas molecules in each chamber. Explain your reasoning and comment on your result.

3. A vessel is constructed to hold small volumes. It is of such a size that it may be considered to only hold a maximum of 24 molecules (*i.e.* a lattice size of 24 sites). Fifteen molecules are introduced into this space. Using appropriate techniques, show that the 15 molecules will use all of the space inside the container, and calculate the statistical entropy for this system.

Solutions to Exercises

Solutions: Exercise 2.1

1. $\Omega = \dfrac{5!}{(3! \times 2!)} = 10$ permutations. Probability $P = \dfrac{10}{2^5} = \dfrac{10}{32}$

2. This is a single permutation, so $\Omega = 1$, $P = \dfrac{1}{2^4} = \dfrac{1}{16}$

3. Equal split, 4H, 4T; $\Omega = \dfrac{8!}{(4! \times 4!)} = 70$, $P = \dfrac{70}{2^8} = \dfrac{70}{256} = \dfrac{35}{128}$

4. $\Omega = \dfrac{4!}{(2! \times 1! \times 1!)} = 12$, $P = \dfrac{12}{6^4} = \dfrac{12}{1296} = \dfrac{1}{108}$

5. $\Omega = \dfrac{4!}{(1! \times 1! \times 1! \times 1!)} = 24$, $P = \dfrac{24}{4^4} = \dfrac{24}{256} = \dfrac{3}{32}$

Solutions: Exercise 2.2

1. $\Omega = \dfrac{12!}{(6! \times 6!)} = 924$; $S = k_B \ln(924) = 6.83\ k_B$

2. $\Omega = \dfrac{12!}{(4! \times 4! \times 4!)} = 34\,650$; $S = k_B \ln(34\,650) = 10.5\ k_B$

3. $\Omega = \dfrac{12!}{(3! \times 3! \times 3! \times 3!)} = 369\,600$; $S = k_B \ln(369\,600) = 12.8\ k_B$

4. $\Omega = \dfrac{10!}{(1! \times 2! \times 3! \times 4!)} = 12\,600$; $S = k_B \ln(12\,600) = 9.44\ k_B$

Solutions: Exercise 2.3

1. $N = \dfrac{40!}{(5! \times 35!)} = 658\,008$, $P = \dfrac{1}{658\,008} = 1.52 \times 10^{-6}$

2. $N = \dfrac{6!}{(4! \times 2!)} = 15$

Solutions: Exercise 2.4

1. NaCl conductivity / mS cm^{-1}: mean, 69.9; mode, 69.5; median, 69.8
2. (−)-menthone boiling point / °C: mean, 206; mode, 205; median, 205
3. $\Delta_c H^{\ominus}$ ethanol / kJ mol^{-1}: mean, 1282; mode, 1211; median, 1285

Solutions: Exercise 2.5

1. $\Delta_{dil}H$ HCl / kJ mol^{-1}: mean = 1.691; s = 0.272
2. Surface tension of H_2O (25 °C) / mN m^{-1}: mean = 75.39; s = 4.01

Solutions: Exercise 2.6

1. Calculate 12 molecules SF_6 in a lattice of 18 sites;
$$\Omega = \frac{18!}{(12! \times 6!)} = 18\,564.$$

2. 4^{12} permutations; multiplicity for equal distribution,
$$\Omega = \frac{12!}{(3! \times 3! \times 3! \times 3!)} = 369\,600,\ P_{equal} = 0.0220\ldots\ \text{Quite a low}$$
probability!

3. Only use 15 lattice sites, $\Omega_{15} = \dfrac{15!}{15!} = 1$;

   expand to 20 lattice sites,
$$\Omega_{20} = \frac{20!}{(15! \times 5!)} = 15\,504;$$
   expand to 24 lattice sites,
$$\Omega_{24} = \frac{24!}{(15! \times 9!)} = 1\,307\,504.$$

Using all the lattice sites gives more possible rearrangements, so a more probable state. S = 14.1 k_B.

Learning Points: What We'll Cover

☐ Precision, accuracy and errors in the context of experimental measurement

☐ The correct use of significant figures in calculating and reporting experimental values

☐ The standard error and its use in combination with the appropriate number of significant figures

☐ Propagating experimental uncertainties through mathematical relationships to report the uncertainty correctly in calculated quantities

☐ Statistical tests to qualify any outliers in a data series and to check whether data sets are statistically different

Why This Chapter Is Important

- In chemistry, we validate our theoretical models through experimental measurement; these measurements are subject to laws of probability and we need to know whether the variations in our measurements are significant.
- Significant figures tell us everything about the measurements made, so it is important that we all use the same rules for reporting these and carrying them through calculations.
- Experimental uncertainty ('experimental error') is unavoidable, so we must be able to quantify 'the bounds of experimental error' in order to allow for them effectively in our analyses.
- We need to make sure that we have a common sense approach to handling errors; if our calculated values have an uncertainty greater than the value itself, we need to be sure that we have handled the errors correctly.

Experimental Uncertainty and Significant Figures: What Are the Bounds of Experimental Error?

It's easy not to think about the number of numbers in a number, but in science the number of significant figures is representative of how well we know a value. Some values we know to a high precision; for example, the speed of light is 299 792 458 m s^{-1}. Thousands of measurements went in to knowing this value so accurately, but we will often approximate it to a value with just one significant figure $(3 \times 10^8$ m s$^{-1})$[†].

In addition to significant figures, quantities based on measurements will also have an uncertainty associated with them. The mass of an electron, m_e, has a value of $9.109\,383\,701\,5(28) \times 10^{-31}$ kg, which we know to have an uncertainty (or error) of $0.000\,000\,002\,8 \times 10^{-31}$ kg.

There are also exact values which should be considered to have an infinite number of significant figures. The gas constant is a recent example of such an 'exact' value; it has a value of 8.314 462 618 153 24 J K^{-1} mol^{-1}.[‡]

Usually, when we report a value, we have given due consideration to how well we can measure that value. This chapter shows you how to determine uncertainties in measured values, and how to calculate the uncertainties in other values derived from your measurements.

> Uncertainties are normally only reported to one significant figure; however, when many hundreds of experiments are undertaken, occasionally a value with more significant figures is used.

3.1 Accuracy and Precision

A fundamental aspect of chemistry is making careful measurements and then using these measurements to calculate further values. In calorimetry, for example, we measure a temperature

---

[†]In 1983 the speed of light was redefined to be exactly 299 792 458 m s^{-1} due to the definition of the metre.

[‡]In 2019, IUPAC redefined the SI base units in terms of defined physical constants, removing the experimental nature of these values. The gas constant, R, is the product of two of these defined constants $(R = N_A k_B)$.

change and from that determine values for the enthalpy of a reaction. The accuracy of our measured temperature change determines how accurately we can know the enthalpy of reaction.

It is not just the precision and scale of our thermometer which limits how well we can determine a value for enthalpy; we also have to consider the random nature of measurements. Making several measurements of any value allows the 'average' of these values to cancel out any noise in the measurements, and repeated measurements improve the confidence in our data.

When considering how we collect our data, we have to be aware of certain terminologies to correctly describe our data and its distribution.

Figure 3.1 shows a collection of target diagrams to introduce and clarify the difference between *accuracy* and *precision* when making measurements, and also to demonstrate the difference between *random* and *systematic* experimental errors.

We all strive for 'good' measurements, but in this context 'good' is an ambiguous word. The words *accurate* and *precise* mean two different things but either an accurate or precise data set may both be described as 'good' in its own way.

The 'accuracy' of a data set is a measure of how close the average is to the 'true' value; however, it says nothing about the set of data itself. Figure 3.1a shows a collection of data points which, when considered 'on average', would look very 'good'. Were this to be an actual data set, the mean of these data is close to the accepted value; however, the spread of the data is poor (it has a large standard deviation; see Section 2.3.2) and, with so few measurements, a single point can skew the 'average' a great deal.

Conversely, the 'precision' is a measure of the 'spread' of our data (the size of the standard deviation). Figure 3.1c shows data with a very high precision; however, the data are clustering around a point away from the centre of the target. These data are precise but not accurate; in statistical terms, the mean is a long way from the accepted value and the standard deviation is small.

3.1.1 Random and Systematic Error

Random errors are a natural part of experimental science; we would think it odd if we were to measure an event many thousands of times and the recorded value was always the same. Instead, we would expect the data to collect around a mean value and display a normal distribution (see Figure 3.3; also see Section 2.3.2). For data which give a good representation of the mean, it may be that the

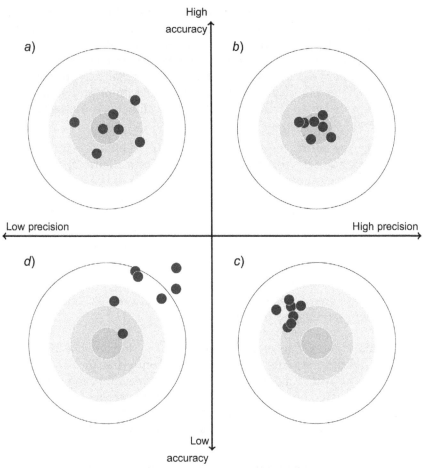

Figure 3.1   Accurate and precise measurements can be visualised by imagining a target; *precise* measurement means measurements which can be reliably reproduced and display a small spread in value, while *accurate* measurement means measurements which, on average, are close to the 'true' value. Something which is *precise* is not necessarily *accurate*, and *vice versa*.

data cluster close together with a small standard deviation (Figure 3.1b), or have a larger spread while still having a mean value close to the accepted value (Figure 3.1a). In either case, as long as the experiment has been conducted to the best of our ability, these random distributions in the data are normal and their effect can be reduced through repeat measurements.

When collecting experimental data, it is easy to think that collecting a number of values with a small spread must mean that the data are accurate. However, we see from Figure 3.1c that this may not be the case. Rather, the deviation from the centre of the target is representative of it giving a value other than the predicted value (it is precise, but not accurate). This is likely to be due to a systematic error—something in our experimental design or in the equipment we are using which biases the results. Typical sources of such errors may include a piece of equipment which has been calibrated incorrectly or an incorrect scaling factor in the analysis stage.

One key source of systematic error is 'rounding errors'. When reading a scale (for example, a volume from a burette), we should round the value to the nearest graduation on the burette. However, there will always be a possibility that the meniscus will lie in such a position that we cannot tell with any certainty whether to round up to the higher graduation or to round down to the lower graduation. In these instances, the convention is to *round to even*:

- If the meniscus lies exactly midway between 41.4 cm^3 and 41.5 cm^3, we *round down* to 41.4 cm^3.
- If the meniscus lies exactly midway between 37.7 cm^3 and 37.8 cm^3, we *round up* to 37.8 cm^3.

It is impossible to read any instrumentation to a greater precision than its scale divisions. Students are often told that "half a scale division is OK", but this cannot always be relied upon, particularly when the indicator is approximately midway between divisions. This is why we use the principle of *round to even* to eliminate any bias when making such measurements.

This ensures that we do not introduce a systematic bias by always rounding mid-range values up (or down!)—statistically, half the time we will round up and half the time we will round down. The same is true of calculated values, when we are reducing them to the appropriate number of significant figures (see Section 3.2). If we have decided that only two significant figures should be given, then:

- If the calculated value is 4.657, this will *round up* to 4.7.
- If the calculated value is 4.642, this will *round down* to 4.6.
- If the calculated value is *exactly* 4.65, this will *round down to even* to 4.6.
- If the calculated value is *exactly* 4.75, this will *round up to even* to 4.8.

3.2 Managing Significant Figures

In our experiments, the significance of our measurements is controlled by the instrumentation and equipment we use in the laboratory; we cover these in Examples 3.2 and 3.3. However, in calculations, a calculator will often return a large number of digits, filling the display. We therefore need to know how to recognise which figures are significant and which are not. The rules for establishing 'significance' are listed in Key Concept 3.1; these will be used throughout the text and we will build on these as we discuss experimental uncertainties and their effect on the precision of measurement.

❗ Key Concept 3.1 When Is a Figure Significant?

$$\Delta_r H = 0.004\,386\,002\,968 \text{ J mol}^{-1} \qquad (3.1)$$

In a value such as the calculated energy shown in eqn (3.1), the significant figures are determined as follows:

- Any non-zero digit is significant.
- Any zero following a non-zero digit is significant.
- Any zero to the left of the first non-zero digit is not significant, *i.e.* the first three zeros in 0.0043... are not significant.
- Zeros to the left of the first non-zero digit simply indicate the position of the decimal point.
- Zeros which are deemed significant must be included to show precision; *e.g.* the following values are all to four significant figures: 7.000, 3000, 45.00.
- Standard form (scientific notation) should be used to avoid ambiguity in significant figures. *e.g.* the value '400 cm' is reported to three significant figures, with an implicit uncertainty of ±1 cm; to report this to two significant figures with an uncertainty of ±10 cm, we need to use standard form: '4.0×10^2 cm'.

There is no 'rule' for how many significant figures to report in a calculation; the appropriate significant figures for the calculated quantity will depend on the precision of the values used in the calculation.

Many students will simply report values to '3 s.f.'; however, you should always appraise your values to ensure that you are using the *appropriate* number of significant figures.

When everything we know is based around taking measurements, it is important to understand how well we can actually 'know' any calculated value, given that this is affected by the precision and accuracy of the readings it is based on. A common experiment done in schools is to calculate the speed of sound; the method for this is summarised in Example 3.1.

Example 3.1 A Simple Experiment to Calculate the Speed of Sound

To calculate the speed of sound:

- Stand a known distance from a wall and make a short loud noise (something like a loud clap is perfect).
- Measure how long it takes to hear the echo.
- Knowing the length of time waited for the sound to return and the distance from the wall (and knowing that the sound has to go to the wall and back again), we can quickly and easily calculate a value for the speed of sound.

Typically, this would determine the speed of sound to be around 300 m s^{-1}. This is a fairly good result (an accepted value is 343 m s^{-1}), but there is ambiguity in this value; not least introduced by the reaction times of the observers.

It is often tempting to simply truncate a value to three significant figures, labelling it '3 s.f.' However, not only are you showing that you have not given due consideration to the appropriate precision, you also assume that your reader cannot count to 'three'! Identify how many significant figures are appropriate, and then just use that many!

We need to remember that zero is still a 'significant figure' and that '300' is no more a rounded figure than '343'. This is why representing values in standard form is so important when communicating appropriate precision, so in this experiment 3×10^2 m s^{-1} or 0.3 km s^{-1} would be the most rigorous way to present this result.

As there is only one significant figure in this reported answer, the 'last' number is also the 'first' number.

Usually, when scientists report a value to a given number of significant figures, the last digit which appears is where we consider any uncertainty to be. When the accepted value for the speed of sound at sea level on a warm day (20 °C) is given as 343 m s^{-1}, it is implied that the level of rigour in the experiment allows this value to be known within 1 m s^{-1}. If there is any uncertainty in the value, it occurs in the final digit; by reporting the value as 343 m s^{-1}, the experimental team are saying, 'The value could be 342 m s^{-1} or it could be 344 m s^{-1} but we are confident that it is very close to 343 m s^{-1}. The value is reported to three significant figures, and if there is any uncertainty it is in the third of these digits.'

Let's now return to the experiment in Example 3.1 and examine the value given there. Is it reasonable to say that we are happy with 300 m s^{-1} to the same level of precision? If the value is written as 300 m s^{-1}, then we are still reporting it to the same number of significant figures.

It is therefore good practice to use either prefixes or standard form (Key Concept 1.9) to show the value to the level of precision we think is reasonable. Our value for the speed of sound becomes 3×10^2 m s^{-1} or 0.3 km s^{-1}, telling the reader that we are pretty happy that it is about 300 m s^{-1}; it might be 200 m s^{-1} or it might be 400 m s^{-1}, but we are happy that if there is an error in a value it is in this first value. Consequently, when we report values from our experiments in chemistry, it is important to be mindful of reporting zeros after decimal points (which may not appear on your calculator); the presence of any zeros tells the reader our confidence in the value we have determined.

Consider the process of preparing an aqueous solution of copper sulfate; we have to measure a mass of the solid $CuSO_4$ and use a balance with its own precision. (Some balances are precise to two decimal places, others to four or even five! See Example 3.2.) We then measure a volume of water and, again, we have a wealth of fluid-handling techniques, each of which has its own precision (Example 3.3). If we are reporting the concentration of our copper sulfate solution to appropriate precision, we need to think about the precision of the measurements we make.

Another example is the ruler in your pencil case compared with a metre rule available in a laboratory. Some rulers have very few divisions (every 1 cm or maybe 0.5 cm); others have more (every 1 mm or even every 0.5 mm). If you were to measure a length using one of these rulers, it should be obvious that you can measure to a higher precision using the ruler with more divisions (Figure 3.2).

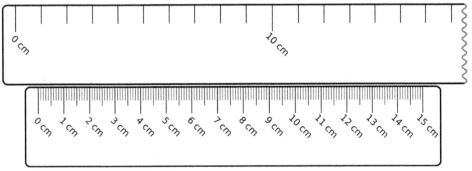

Figure 3.2   A metre rule (top) tends to have fewer divisions than the rule you may have in your own stationery set (bottom). This gives your ruler a higher precision, albeit with a more limited range.

Example 3.2 Using Significant Figures: Balances

Two types of balance are commonly used in chemistry laboratories. Two decimal place (d.p.) balances for rough masses (frequently used in synthetic laboratories, where high precision is not necessary) and four decimal place analytical balances for either smaller masses or increased precision, as required in analytical chemistry.

A mass of 0.4 g of dry $CaCO_3$ was weighed on each of these balances.

- *2 d.p. balance.* This balance weighs 0.40 g $CaCO_3$. If there is an uncertainty in this value, it is in the last digit; therefore, we know for sure that the 'true' mass lies somewhere between 0.39 and 0.41 g.
- *4 d.p. balance.* This balance weighs 0.4000 g $CaCO_3$. Again, any uncertainty lies in the last digit; this time, we know the mass more precisely than before, with the 'true' mass lying somewhere between 0.3999 and 0.4001 g.

Example 3.3 Using Significant Figures: Glassware

- *50 mL burette.* These typically have graduations every 0.1 mL; therefore, we can only know a value to 1 d.p. (±0.1 mL). It is not possible to be any more precise; if the meniscus is between two graduations we must pick the one to which it is closer; if we cannot tell, we round to the even value.
- *25 mL measuring cylinder.* These typically have graduations every 1 mL. Therefore, we can only read the scale to a limit of ±1 mL.

We call this process of rounding 'rounding to even'. The reason for this is that if we always 'round up', we introduce a small bias to our end value. By 'rounding to even', statistically, half the time we will 'round up' to the even value, and half the time we will 'round down' to the even value (see Section 3.1.1).

❓Exercise 3.1 Significant Figures

Determine the number of significant figures in each of the following numbers:

1. 1.3 m
2. 0.0135 g
3. 298.15 K
4. 50.00 mL

5. $9.109\,383\,701\,5 \times 10^{-31}$ kg
6. $0.001\,005\,001\,5$
7. -393.51 kJ mol^{-1}
8. Exactly 1 L

3.2.1 Combining Significant Figures

Having collected the appropriate significant figures in the laboratory, we need to know how to feed these through all of our calculations to report our results to the *appropriate* precision. This is a process called *error propagation*. The overriding principle in error propagation is 'The precision of the result is always limited by the least precise value.'

The exception to this principle is when 'exact' values appear in equations. Examples of these include molecular formulae or integer scaling values in equations. In determining the molecular mass of the allotrope of sulfur S_8, there are *exactly* eight sulfur atoms in a molecule. Although determining the molecular mass of S_8 appears to be a multiplication, it is formally just lots of addition steps and so the addition rule should be followed (*i.e.*, keep the decimal places, as shown in eqn (3.2)):

$$M_r = 8 \times 32.06 \text{ g mol}^{-1} = 256.48 \text{ g mol}^{-1} \qquad (3.2)$$

Likewise, if a 'number' appears in an equation, it does not limit the number of significant figures. For example, in eqn (3.3), both the 8 and the 3 are integers and so may be considered to have 'infinite significant figures'. Consequently, when we evaluate a diffusion-controlled rate constant in eqn (3.4), we disregard any significant figures attached to the '8' or the '3' and just focus on the measurements going into the calculation:

$$k_d = \frac{8RT}{3\eta} \qquad (3.3)$$

$$k_d = \frac{8 \times 8.31446\,\text{J K}^{-1}\,\text{mol}^{-1} \times 293.2\,\text{K}}{3 \times 0.5940 \times 10^{-3}\,\text{Pa s}} = 1.094 \times 10^7\,\text{m}^3\,\text{mol}^{-1}\,\text{s}^{-1} \quad (3.4)$$

Unit conversions are also 'exact' so there is no need to change the number of significant figures.

One other place where this 'just a number' comes up is in electrochemistry; for example, in the Nernst equation, the variable n is the number of electrons transferred in the process and so can only take integer values (with infinite significant figures!).

The solution to eqn (3.4) has four significant figures because it is a multiplication and division and so we preserve the fewest significant figures in the calculation—see Key Concept 3.2.

We cover the rules for propagating errors through different mathematical functions in Key Concepts 3.5, 3.6, 3.7, 3.8 and 3.9. It is not the purpose of this text to show the derivation of all of these rules as there are many excellent texts which illustrate them (see our Reading List at the end of this book for suggestions). For now, all we need is to be able to apply the error propagation correctly so that we can report our results clearly and to the appropriate precision.

---

❶ Key Concept 3.2 The Use of Significant Figures When Adding or Subtracting Values

When adding or subtracting values, the result should preserve the *decimal places* of the *least precise* initial value.

For example, if a temperature is reported to be 20 °C, and you wish to convert this to kelvin, the formula is:

$$T \text{ (in K)} = T \text{ (in °C)} + 273.15 \text{ K}$$

So for our example with 20 °C:

$$T \text{ (in K)} = 20 \text{ °C} + 273.15 \text{ K}$$

$$T = 293 \text{ K}$$

This is because the 20 °C has no decimal places, and the conversion factor has two decimal places. In line with the rule, we preserve the decimal places of the *least precise* value, so we report no decimal places in our result.

Common calculations in chemistry in which values are added or subtracted are:

1. Molecular masses
2. Temperature conversions
3. Enthalpies of reaction
4. Entropies of reaction
5. Cell potentials

❓Exercise 3.2 Significant Figures in Addition and Subtraction

For each of the following, evaluate to the correct number of significant figures.

1. $T \text{ (in K)} = 25.0 \text{ °C} + 273.15 \text{ K}$
2. $M_r = 15.999 \text{ g mol}^{-1} + 1.0079 \text{ g mol}^{-1} + 22.99 \text{ g mol}^{-1}$
3. $E_{\text{cell}} = 0.332 \text{ V} - (-0.7628 \text{ V})$
4. $\Lambda_{\text{NaOH}} = 5.011 \text{ S cm}^2 \text{ mol}^{-1} + 19.8 \text{ S cm}^2 \text{ mol}^{-1}$
5. $\Delta_c H_{\text{diamond}} = -393.51 \text{ kJ mol}^{-1} - 1.895 \text{ kJ mol}^{-1}$
6. $\Delta c_{p,\text{m}} = 35.31 \text{ J K}^{-1} \text{ mol}^{-1} - (28.824 \text{ J K}^{-1} \text{ mol}^{-1} + 28.824 \text{ J K}^{-1} \text{ mol}^{-1} + 8.527 \text{ J K}^{-1} \text{ mol}^{-1})$
7. $\Delta H = -38.64 \text{ kJ mol}^{-1} + 965 \text{ J mol}^{-1}$
8. $p = 101.325 \text{ k Pa} + 1742 \text{ Pa}$

When something is described as being 'exactly' a given value, we can consider this as having an infinite number of significant figures.

> **❶ Key Concept 3.3 The Use of Significant Figures When Multiplying or Dividing Values**
>
> When multiplying or dividing values, the result should preserve the *significant figures* of the *least precise* initial value.
>
> For example, if we are to determine the number of moles in a sample of benzoic acid ($M_r = 122.122$ g mol^{-1}) of mass 0.1654 g, then:
>
> $$n = \frac{m}{M_r}$$
>
> $$n = \frac{0.1654\,g}{122.122\ g\ mol^{-1}} = 1.354 \times 10^{-3}\ mol$$

❓Exercise 3.3 Significant Figures in Multiplication and Division

For each of the following, evaluate to the correct number of significant figures.

1. $q = kT = 1.45$ kJ K$^{-1} \times 2.3$ K

2. $K = \dfrac{[A]}{[B]} = \dfrac{1.5982}{0.63}$

3. $m = nM_r = 25.0$ mol $\times 46.069$ g mol^{-1}

4. $p = \dfrac{nRT}{V} = \dfrac{1.04\ mol \times 8.31446\ J K^{-1}\ mol^{-1} \times 340\ K}{1.406\ dm^3}$

5. $k_d = \dfrac{8RT}{3\eta} = \dfrac{8 \times 8.31446\ J K^{-1}\ mol^{-1} \times 298.15\ K}{3 \times 0.8900 \times 10^{-3}\ Pa\,s}$

6. $c = \dfrac{m}{M_r v} = \dfrac{1.4528\ g}{58.44\ g\,mol^{-1} \times 100.00\ mL}$

7. $m = \rho v = 0.789$ g cm$^{-3} \times 50.00$ cm^3

8. $n = cv = 150$ mmol dm$^{-3} \times 100.00$ cm^3

Remember to cancel the units in this equation (Question 4); the volume must be in m³!

Think about changing the prefix in units so that they can cancel (Question 8).

❓Exercise 3.4 Determining Molecular Masses to the Appropriate Significant Figures

Given the following values of M_r, determine the relative molecular mass to the appropriate number of significant figures, and the number of moles, given the provided masses in each case.

Element	M_r / g mol^{-1}
H	1.0079
C	12.011
N	14.007
O	15.999
Na	22.990

Element	M_r / g mol^{-1}
S	32.06
Cl	35.45
Ru	101.07

1. NaCl, 1.45 g	5. C_8H_6S, 10.40 g
2. CH_3CH_2OH, 2.3457 g	6. $C_{16}H_{18}ClN_3S$, 0.0039 g
3. $HOOCCH_2CH_2COOH$, 3.22 g	7. $C_{28}H_{31}N_2O_3Cl$, 83.5 mg
4. C_6H_5COONa, 0.0587 g	8. $[Ru(C_{10}H_8N_2)_3]^{2+}$, 7.9 mg

When taking a logarithm, the 'rule of thumb' is to gain a significant figure, but there is actually hidden detail here (Key Concept 3.4). Logs are common in chemistry, with natural logs (ln) appearing in a huge range of places, therefore being able to manipulate significant figures quickly when using logs is incredibly useful.

When determining the number of significant figures in a calculation with both logs and multiplication, you should follow the priority rules of BODMAS (or BIDMAS); see Toolkit A.1.

---

❶ Key Concept 3.4 Rules of Significant Figures With Logarithms and Indices

Determining the number of significant figures required when taking a logarithm requires some of the 'rules of logs' laid out in Key Concept 1.5.

Start by expressing the value in terms of scientific notation, for example 0.000 041 8 becomes 4.18×10^{-5}. When taking the log of this:

$$\log (4.18 \times 10^{-5}) = \log 4.18 + \log 10^{-5} \qquad (3.5)$$

We can consider the $\log 10^{-5}$ term, when calculated, to have an infinite number of significant figures because here both 10 and -5 are integer values.

For the $\log 4.18$ term, the correct number of significant figures would be the same as the input value (or we could consider this to be increasing the number of decimal places by 1), so:

$$\log 4.18 = 0.621 \qquad (3.6)$$

When combining these terms:

$$\log (4.18 \times 10^{-5}) = 0.621 - 5.000\ 00... = -4.379 \qquad (3.7)$$

In effect, we have gained a significant figure.[§]

When taking a power or index you should (usually) lose a significant figure; this means that if you take a log and then an index of resultant value, you should end up with the same number of significant figures that you started with.

The same would be true if you had log (4.18 × 10^5)—you would gain a significant figure.

[§]If the value is between 1 and 10, then we should not gain a significant figure, as when we take the logarithm of 10^0, we obtain a value of zero with no decimal places!

When taking a log('log to the base ten'), if the value lies between 10^0 and 10^1, we *do not* gain a significant figure. When taking a ln ('log to the base e'), if the value lies between e^0 and e^1, we *do not* gain a significant figure.

Suppose we were to determine a value of the driving force, ΔG, of a reaction using:

$$\Delta G = -RT \ln K \tag{3.8}$$

If $K = 4.1$, then at 298.15 K, $\Delta G = -3.49$ kJ mol^{-1} (we have gained a significant figure). However, if $K = 2.1$, then at 298.15 K, $\Delta G = -1.8$ kJ mol^{-1} because here we do not gain a significant figure, as the value of K is between e^0 and e^1.

Example 3.4 Significant Figures in Calculations

N_2O_5 is in equilibrium with its decomposition products NO_2 and NO_3. Given that the concentrations of these at 25.0 °C are 7.4×10^{-9} mol dm^{-3}, 0.61 mol dm^{-3} and 0.42 mol dm^{-3}, respectively, determine the Gibbs free energy for this reaction (eqn (3.8)). You may assume that activities are equal to concentrations, as the concentrations are low:

$$K = \frac{[NO_2][NO_3]}{[N_2O_5]}$$

The first step would be to determine K (the highest priority, as it is essentially in brackets), to the appropriate significant figures:

$$K = \frac{0.61 \times 0.42}{7.4 \times 10^{-9}} = 3.5 \times 10^{10}$$

Then (taking the operators in order of priority), we determine the value for the natural logarithm (remembering that we gain a significant figure as the value within the logarithm is greater than e^1) :

$$\ln K = 24.3$$

The temperature is (remembering to keep to the decimal places of the least precise value):

$$T = (25.0 + 273.15) \, K = 298.2 \, K$$

Finally, the rest of the calculation can be performed, knowing the appropriate numbers of significant figures for each component:

$$\Delta G = -8.31446 \text{ J K}^{-1} \text{ mol}^{-1} \times 298.2 \text{ K} \times 24.3$$

$$\Delta G = -60.2 \text{ kJ mol}^{-1}$$

3.3 The Standard Error

There is always random error in any measurement, and this random error follows a particular distribution around the 'actual' value. This distribution is called a normal, or Gaussian, distribution (Figure 3.3). For a given method of measurement, you would not expect the shape of the distribution (and consequently the 'spread' of the distribution) to change, no matter how many measurements you take. The *sample* standard deviation, s, for a data set (which is a measure of this spread) will remain the same. We

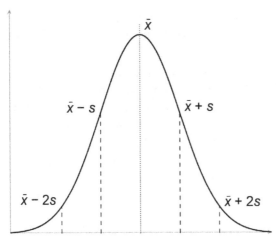

Figure 3.3 Experimental data will naturally include random error. The data will collect with a frequency of appearance which follows a Gaussian distribution, with the first and second standard deviations marked from the mean.

introduced the concept of data distributions in Chapter 2, when looking at statistics; however, it is now time to look at how this may be used to determine the standard error in our measurements.

Normally we would expect to be more certain of a value the more measurements we take. This is reflected in eqn (3.9), where $\sigma_{\bar{x}}$ is the standard error in the mean value of x, s the standard deviation (usually the sample standard deviation) and N is the number of samples taken. Clearly, the more measurements we take, the larger the value of $\sqrt{N}$ and the smaller the standard error becomes:

$$\sigma_{\bar{x}} = \frac{s}{\sqrt{N}} \tag{3.9}$$

The standard error should normally only be reported to one significant figure. This is true unless very many measurements have been made. (In some experiments, thousands of measurements are taken to report standard errors to two or three significant figures!) The standard error should also always be rounded up— as we are trying to account for an uncertainty, we do not want to underestimate this.

Standard errors are reported in one of two common forms. The first uses a ± notation to indicate the error:

$$\bar{x} \pm \sigma_{\bar{x}} \tag{3.10}$$

In this notation, the standard error should appear as its full value, for example: $T = 34.28 \pm 0.07$ K.

The second method uses brackets to indicate the standard error

$$\bar{x}(\sigma_{\bar{x}}) \tag{3.11}$$

The standard error is also sometimes called the 'standard deviation of the mean' (SDOM); the reduction of the standard error as we take more samples shows that we have a more accurate estimate of the mean.

Standard errors are normally only reported to one significant figure because the standard error itself has an uncertainty and there is no point in being overly 'accurate'; both the mean and its uncertainty have an 'error'. See I. G. Hughes and T. P. A. Hase, *Measurements and Their Uncertainties: A Practical Guide to Modern Error Analysis*, OUP, Oxford, 2010.

The standard error is just an estimate of the error in the measurement. Like the mean, as the number of measurements increases, our confidence in the value of standard error increases.

You will find the terms 'error' and 'uncertainty' used interchangeably—they are both referring to the standard error

The same example, in bracket notation, is reported as $T = 34.28(7)$ K, the error (or uncertainty) appears in the number immediately before the bracket.

Each of these methods is reporting the same value and uncertainty, and each can be seen used in books and journals. You will find that the bracket notation is more commonly used as it is easier to read with many decimal places; as a result, it is the version which we will use predominantly in this text.

In calculating a standard error, you may find that it is larger than the number of significant figures in the mean. In this case, you will need to reduce the number of significant figures in a calculated value of the mean because the standard error is not occurring in the last significant figure. An example of this is shown in Example 3.5, part 2.

Example 3.5 Calculating a Standard Error

1. A kinetics experiment is conducted in which the initial rate of reaction was recorded five times:

$$\frac{d[A]}{dt} / 10^{-4} \text{ mol dm}^{-3} \text{ s}^{-1}: 1.24, 1.45, 1.36, 1.27, 1.27$$

The mean of the values is calculated as 1.32×10^{-4} mol dm^{-3} s^{-1}.

The sample standard deviation can be calculated *via*:

$$s = \sqrt{\frac{1}{N-1}\Sigma\left(x - \overline{x}\right)^2}$$

$$s = \sqrt{\frac{1}{5-1} \times 29.9 \times 10^{-3}} = 0.09 \times 10^{-4} \text{ mol dm}^{-3} \text{ s}^{-1}$$

And the standard error can be calculated as:

$$\sigma_{\overline{\text{rate}}} = \frac{s}{\sqrt{N}}$$

$$\sigma_{\overline{\text{rate}}} = \frac{0.09}{\sqrt{5}} = 0.05 \times 10^{-4} \text{ mol dm}^{-3} \text{ s}^{-1}$$

This value of 0.05 has been rounded up from 0.0402 ... because you should never underestimate the error in a value.

The rate of reaction can either be reported as $1.32 \pm 0.05 \times 10^{-4}$ mol dm^{-3} s^{-1} or in the bracket form of $1.32(5) \times 10^{-4}$ mol dm^{-3} s^{-1}.

2. The experiment is repeated; this time the following data are obtained:

$$\frac{d[A]}{dt}/10^{-4} \text{ mol dm}^{-3} \text{ s}^{-1}: 1.42, 1.12, 1.69, 1.61, 1.59$$

We now calculate the mean of the values and the standard error:

$$\overline{\text{rate}} = 1.49 \times 10^{-4} \text{ mol dm}^{-3}$$

$$\sigma_{\overline{\text{rate}}} = 0.1 \times 10^{-4} \text{ mol dm}^{-3}$$

In this situation, we see that the standard error is now larger than the last significant figure in the mean value; this means that we need to reduce the number of significant figures in reporting the mean value, as its precision is greater than the uncertainty in the measurement. Our final value is then $1.5(1) \times 10^{-4}$ mol dm^{-3} s^{-1} (or $1.5 \pm 0.1 \times 10^{-4}$ mol dm^{-3} s^{-1}).

❓Exercise 3.5 Determining Standard Errors

Determine the mean and standard error for each of the following data sets, and report in an appropriate single-answer form.

1. A: 0.074, 0.081, 0.075, 0.071, 0.084, 0.084, 0.072
2. p/mm Hg: 131, 145, 139, 147, 143, 130, 147
3. I/mA: 1.49, 1.42, 1.47, 1.42, 1.45
4. V/V: 2.82, 2.81, 2.84, 2.81, 2.85, 2.82
5. ΔT/K: 2.42, 2.33, 2.18, 2.49, 2.35, 2.45
6. m/g: 1.0026, 1.0019, 0.9972, 0.9986, 1.0009
7. T/°C: 32.48, 31.95, 32.25, 31.98, 32.34
8. ΔH/kJ mol^{-1}: −101.4, −109.2, −99.5, −113.7, −108.6, −100.4, −118.0, −111.1[¶]

> Use a spreadsheet or other computer-based method for these calculations.

3.4 Combining Uncertainties

We saw in Section 3.2.1 how we can combine measurements with different *significant figures*, but we now need to find how to combine the determined uncertainty in our measurements. The origin of these equations is not discussed in this text;[l] rather we wish to cover the basics of 'error analysis', which is usually all a chemist needs for practical studies.

Each mathematical function we use in processing our data requires a different method to propagate the error. This will often require several propagation steps to quantify the final uncertainty in our

> It is always worth paying attention to the magnitude of the uncertainty; if you reach a point where your uncertainty is larger than the value itself (e.g. a value of 3.45 ± 4 m), you may need to revisit the error propagation.

[¶] These are real data from an experiment in which eight groups of students calculated an enthalpy of reaction.

[l] The book *An Introduction to Error Analysis* (J. R. Taylor, University Science Books, Mill Valley, 2nd edition, 1997) gives an excellent introduction to the origin of the following equations. An alternative text, *Measurements and their Uncertainties* (I. G. Hughes and T. P. A. Hase, OUP, Oxford, 2010) is another excellent reference.

Formally, we can combine all errors in a single step, regardless of the mathematical operation used; see Toolkit A.8.

calculated quantity. As before, these have different 'rules', depending on whether quantities are added or subtracted, multiplied or divided, and so on.

Additionally, the standard error (Section 3.3) of a value may be used to further propagate errors in values calculated from it.

❗ Key Concept 3.5 Uncertainties in Terms which are Added or Subtracted

When adding or subtracting values, such as:

$$a = x + y - z \ldots \quad (3.12)$$

where one or more of the values has an associated uncertainty (δx, δy, etc.), the overall uncertainty in a can be found as:

$$\delta a = \sqrt{\delta x^2 + \delta y^2 + \delta z^2} \ldots \quad (3.13)$$

For practical purposes, the simplified version of the equation can be used:

$$\delta a \approx \delta x + \delta y + \delta z \ldots \quad (3.14)$$

where δa is the uncertainty of the calculated value, δx, δy, and δz are the uncertainties in our measured values.

It doesn't matter if the variables are added or subtracted; the uncertainty equation is still the same.

Uncertainties are *always* additive, even if terms are subtracted or divided.

Example 3.6 Combining Uncertainties which are Added or Subtracted

Q: In an experiment to determine the enthalpy of dissolution of two salts of sodium carbonate, Na_2CO_3 and $Na_2CO_3 \cdot (H_2O)_{10}$, the values were determined to be $-23.5(8)$ kJ mol^{-1} and $+68.0(9)$ kJ mol^{-1} respectively. Determine the enthalpy of hydration of sodium carbonate.

A: We can break down the problem into the steps we will follow.

• Determine the balanced chemical equation for the reaction.
• Determine the enthalpy of reaction for the hydration.
• Work out the uncertainty in this value.

We start by writing out the chemical reaction for the hydration of sodium carbonate:

$$Na_2CO_3 + 10H_2O \rightarrow Na_2CO_3 \cdot (H_2O)_{10}$$

The enthalpy of hydration may be determined because you know the enthalpy of dissolution of the two states. So *via* Hess's law:

$$\Delta_{hyd}H_{Na_2CO_3} = \Delta_{diss}H_{Na_2CO_3} - \Delta_{diss}H_{Na_2CO_3 \cdot (H_2O)_{10}}$$

$\Delta_{hyd}H_{Na_2CO_3} = -23.5$ kJ mol^{-1} $- 68.0$ kJ mol^{-1} $= -91.5$ kJ mol^{-1}

Remember from Key Concept 3.5:

$$\delta a \approx \delta x + \delta y + \delta z...$$

$$\delta\Delta_{hyd}H_{Na_2CO_3} \approx \delta\Delta_{diss}H_{Na_2CO_3} + \delta\Delta_{diss}H_{Na_2CO_3\cdot(H_2O)_{10}}$$

$$\delta\Delta_{hyd}H_{Na_2CO_3} \approx 0.8 \text{ kJ mol}^{-1} + 0.9 \text{ kJ mol}^{-1} = 1.7 \text{ kJ mol}^{-1}$$

which is rounded up to one significant figure of 2 kJ mol^{-1}. So we have $\Delta_{hyd}H_{Na_2CO_3} = -92(2)$ kJ mol^{-1}.

For context, using the full version of the equation:

$$\delta\Delta_{hyd}H_{Na_2CO_3} = \sqrt{\delta\Delta_{diss}H^2_{Na_2CO_3} + \delta\Delta_{diss}H^2_{Na_2CO_3\cdot(H_2O)_{10}}}$$

$$\delta\Delta_{hyd}H_{Na_2CO_3} = \sqrt{0.8^2 + 0.9^2} = 1.2 \text{ kJ mol}^{-1}$$

which still rounds up to one significant figure and so gives the same result.

❓Exercise 3.6 Propagating Errors—Addition and Subtraction

Determine the value and uncertainty in the subject of each of the following equations:

1. $\Delta T = T_2 - T_1$; $T_1 = 25.4(2)$ °C and $T_2 = 21.4(2)$ °C
2. $m = m_2 - m_1$; $m_2 = 6.4582(1)$ g and $m_1 = 6.2323(1)$ g
3. $\Delta_r H = \Delta_f H_1 - \Delta_f H_2$; $\Delta_f H_1 = +35(3)$ kJ mol^{-1} and $\Delta_f H_2 = -78.2(8)$ kJ mol^{-1}
4. $\Delta C_p = (C_{p,H_2O} + C_{p,H_2O}) - (C_{p,H_2} + C_{p,H_2} + C_{p,O_2})$; $C_{p,H_2O} = 75.29(8)$ J K^{-1} mol^{-1}, $C_{p,H_2} = 28.82(3)$ J K^{-1} mol^{-1} and $C_{p,O_2} = 29.4(4)$ J K^{-1} mol^{-1}

❗ Key Concept 3.6 Uncertainties in Terms which are Multiplied or Divided

When multiplying or dividing two values (x, y, z etc.) which have associated uncertainties (δx, δy, etc.), the overall uncertainty in the calculated value a, δa, can be found as:

$$\frac{\delta a}{a} = \sqrt{\left(\frac{\delta x}{x}\right)^2 + \left(\frac{\delta y}{y}\right)^2 + \left(\frac{\delta z}{z}\right)^2 + \cdots}$$

$$(3.15)$$

The vertical lines around a value are called *modulus* signs; they mean that you should ignore the sign of the value (whether positive or negative) and just consider the magnitude of the value.

For the units to cancel in the ideal gas equation, volumes must be in m³, pressures in pascals and temperatures in kelvin.

The diameter term appears twice as it is squared in the original calculation.

The uncertainties need to be propagated through the unit conversions; sometimes, this just means 'ignoring' the 'exact' unit conversion terms.

Again there is a simpler version of this equation, which will give an approximate value of the uncertainty:

$$\frac{\delta a}{|a|} \approx \frac{\delta x}{|x|} + \frac{\delta y}{|y|} + \frac{\delta z}{|z|} \cdots \tag{3.16}$$

If the value with the uncertainty, x (with uncertainty δx), is multiplied by a constant, or by a value that is exact (k):

$$a = kx$$

then:

$$\delta a = |k|\delta x \tag{3.17}$$

An example of this would be converting the pressure in bar to the pressure in pascals. If the pressure in bar, $p_{bar} = 2.4(2)$ bar, then using eqn (3.17), $p_{Pa} = 2.4(2) \times 10^5$ Pa.

Example 3.7 Determining the Uncertainty in the Ideal Gas Equation

Q: Determine the uncertainty in the number of moles of gas in a cylindrical vessel 20.0(5) cm in diameter and 50.0(5) cm long; pressure, p, = 4.52(9) bar; temperature, $T = 25.3(9)$ °C.

A: Break down the problem into its constituent parts:

• Determine the number of moles of gas, remembering to use appropriate units so that they cancel.
• Work out the uncertainty in this value.

Firstly, calculate the value of n:

$$n = \frac{pV}{RT} = \frac{4.52 \times 10^5 \text{ Pa} \times \left(\frac{\pi}{4} \times 20.0^2 \times 50.0\right) \times 10^{-6} \text{ m}^3}{8.314\,462 \text{ J K}^{-1} \text{ mol}^{-1} \times (25.3 + 273.15) \text{ K}}$$

$$n = 2.86 \text{ mol}$$

Now the uncertainty may be calculated. We could first work out an uncertainty in a volume, but it is simpler to just consider the whole equation at once, as we did when calculating the volume:

$$\frac{\delta n}{n} \approx \frac{\delta p}{p} + \frac{\delta d}{d} + \frac{\delta d}{d} + \frac{\delta l}{l} + \frac{\delta T}{T}$$

$$\frac{\delta n}{2.86} \approx \frac{0.09}{4.52} + \frac{0.5}{20.0} + \frac{0.5}{20.0} + \frac{0.5}{50.0} + \frac{0.9}{298.4}$$

$$\delta n \approx 0.24$$

This rounds up to $\delta n = 0.3$. Since the error is in the first decimal place, the calculated value needs to be rounded to reflect this level of uncertainty. After the calculation, $n = 2.9(3)$ mol.

Exercise 3.7 Propagating Errors—Multiplication and Division

Convert each of the following values and corresponding uncertainties to the specified units:

1. $p = 50.24(8)$ bar; MPa
2. $V = 3.25(4)$ dm^3; m^3
3. $p = 755(5)$ mm Hg; Pa
4. $\lambda = 409(6)$ nm; E in J

> $E = hc/\lambda$.
>
> Both h and c are defined as absolute values and so have no uncertainties associated with them.

Determine the value and uncertainty in the subject of each of the following equations:

5. $C = \dfrac{q}{\Delta T}$; $q = 615(7)$ J and $\Delta T = 1.58(6)$ °C

6. $[\text{NaCl}] = \dfrac{m}{M_r V}$; $M_r = 58.443$ g mol^{-1}, $V = 50.00(2)$ dm^3, $m = 2.5647(1)$ g

7. $n = \dfrac{pV}{RT}$; $p = 5.27(3)$ bar, $V = 2.500(1)$ dm^3 and $T = 23.98(2)$ °C

8. $S = \dfrac{q_{rev}}{nT}$; $q_{rev} = 15.0(3)$ kJ, $n = 2.48(6)$ mol and $T = 273.2(2)$ K

> Remember the unit conversions!

❗ Key Concept 3.7 Uncertainties in Powers (Indices)

When dealing with powers in uncertainty calculations (for example, terms which are squared), where:

$$a = x^n \tag{3.18}$$

then the uncertainty in the value a (δa), may be determined using:

$$\frac{\delta a}{|a|} = \left| n \frac{\delta x}{x} \right| \tag{3.19}$$

This is true no matter whether the power, n, is positive or negative, integer or fractional.

Example 3.8 Determining the Uncertainty in the Kinetic Energy of a Particle

Q: Determine the kinetic energy (and its uncertainty) of molecular fluorine (a F_2 gas molecule) travelling at $6.0(2) \times 10^2$ m s^{-1}.

A: Break down the problem into its constituent parts:

- Determine the mass of a fluorine molecule in the appropriate units.
- Determine the energy of the fluorine molecule.
- Calculate the uncertainty in this value.

Remember that the SI base unit of mass is kg, so we need to carry out this conversion! This should be *kg per mol* to allow the units to cancel.

Expressing our conversion factor as '10^{-3} kg g^{-1}' is saying "There are 0.001 kilograms per gram". This allows for straightforward scaling.

Firstly, determine the mass of a molecule of fluorine (F_2, $M_r = 2 \times 18.998\,403$ g mol^{-1}):

$$m_{F_2} = \frac{M_r}{N_A} \times 10^{-3} \text{ kg g}^{-1} = \frac{2 \times 18.998\,403 \text{ g mol}^{-1}}{6.022\,140\,9 \times 10^{23} \text{ mol}^{-1}} \times 10^{-3} \text{ kg g}^{-1}$$

$$m_{F_2} = 6.309\,5179 \times 10^{-26} \text{ kg}$$

Now we can work out the kinetic energy of a single molecule using:

$$E = \frac{1}{2}mv^2 \tag{3.20}$$

$$E = \frac{1}{2} \times 6.309\,5179 \times 10^{-26} \text{ kg} \times \left(6.0 \times 10^2 \text{ m s}^{-1}\right)^2$$

$$E = 1.1 \times 10^{-20} \text{ J}$$

Finally, we determine the uncertainty in the calculated value of kinetic energy. Since there is only an uncertainty in the velocity term, v, we can use eqn (3.19), where $n = 2$ (the velocity is squared, so its index is equal to 2):

$$\delta E = |E| \times \left| n \frac{\delta v}{v} \right| = 1.1 \times 10^{-20} \text{ J} \times \left(2 \times \frac{0.2 \times 10^2 \text{ m s}^{-1}}{6.0 \times 10^2 \text{ m s}^{-1}} \right)$$

$$\delta E = 0.1 \times 10^{-20} \text{ J}$$

Given the significant figures in the question, the uncertainty rounds up to just one decimal place and we can state that the energy of the particle $E = 1.1(1) \times 10^{-20}$ J.

In calculating the uncertainty when values are added (or subtracted), we can approximate this as the sum of the uncertainties of each of the values (Key Concept 3.5); we can make a similar approximation when multiplying or dividing, but now we approximate the overall fractional uncertainty $\left(\delta a/a\right)$ as the sum of the *fractional error* of each of the values (Key Concept 3.6). When we propagate an uncertainty within a power, we find that the overall fractional error is approximated as the sum of the fractional errors (the equation in Key Concept 3.7 can be derived from the expression in Key Concept 3.6).

When we are using logarithms to calculate any value, we have to operate slightly differently depending on the base of the logarithm. In chemistry, we typically only work with natural logarithms (ln $x \equiv \log_e x$) or base 10 logarithms ($\log_{10} x \equiv \log x$).** The most fundamental example comes from the natural logarithm, so we will start with that function.

_____

**The same principles apply to any other base as to base 10, but no other base is used in chemistry.

> **① Key Concept 3.8 Uncertainties in Logarithms**
>
> When taking the natural logarithm of any function (eqn (3.21)), the error or uncertainty in that value is propagated using eqn (3.22):
>
> $$a = \ln x \qquad (3.21)$$
>
> $$\delta a = \frac{\delta x}{x} \qquad (3.22)$$
>
> We discussed the relationship between different bases of logarithms in Section 1.2.5. The *natural logarithm* is mathematically fundamental; if we have logs to any other base, we need to 'correct' the term by using the *rules of logs*. In chemistry, there is no need for logarithms to any base other than e (the natural log, used in kinetics) or 10 (used in pH notation):
>
> $$a = \log x \qquad (3.23)$$
>
> $$\delta a = \frac{\delta x}{\ln(10)x} \qquad (3.24)$$

When we consider uncertainties in exponentials, it is, once again, most convenient to look at exponentials of e (Key Concept 3.9). As the inverse function of a logarithm, all other exponents are fundamentally linked to this natural exponent in a similar manner to the link between all logarithms and the natural logarithm.

> **① Key Concept 3.9 Uncertainties in Exponentials**
>
> Just like 'natural' and 'base 10' logarithms, the two exponents e^x and 10^x are fundamentally linked:
>
> $$a = e^x \qquad (3.25)$$
>
> $$\delta a = a\delta x \qquad (3.26)$$
>
> and for an exponent of 10:
>
> $$a = 10^x \qquad (3.27)$$
>
> $$\delta a = \ln(10) \times a\delta x \qquad (3.28)$$
>
> The uncertainty for a power of 10 is simply scaled by the factor of $\ln(10) \approx 2.302...$ relative to the uncertainty for a natural exponential, e^x.

Remember that an *exact* value is considered to have an infinite number of decimal places.

V_f stands for *final* volume and V_i stands for *initial* volume.

💬

Work done being negative means that the system is losing energy; *i.e. doing work* on the surroundings, If this were positive, the system would be *gaining energy* as the surroundings do work on the system. Therefore, it is really important to consider the sign during this calculation.

If you handle the units in this way, you see that the units of volume cancel and you don't need to worry about converting them.

Remember that you should gain a significant figure when using a natural logarithm.

Example 3.9 Determining the Uncertainty in Work Done by an Expanding Gas

Q: Determine the work done, w, and its associated error in a reversible expansion of a gas, when[a] exactly 2 moles of a gas increases in volume from 2.1(2) dm^3 to 6.4(2) dm^3 at 298.2(2) K:

$$w = -nRT \ln \frac{V_f}{V_i} \qquad (3.29)$$

A: As always, if you break down the problem into parts it becomes more manageable.

● Determine the ratio of volumes, and hence the work done.
● Determine the uncertainty in the ratio of the volumes.
● Propagate this uncertainty into the work done.

The work done is calculated using eqn (3.29):

$$w = -2 \text{ mol} \times 8.314\,46 \text{ J K}^{-1} \text{ mol}^{-1} \times 298.2 \text{ K} \times \ln\left(\frac{6.4 \text{ dm}^3}{2.1 \text{ dm}^3}\right)$$

$$\frac{V_f}{V_i} = x = 3.0$$

and:

$$w = -5.53 \text{ kJ}$$

Now we can propagate the error (note the shorthand, $V_f/V_i = x$):

$$\delta_x = x\left(\frac{\delta V_f}{V_f} + \frac{\delta V_i}{V_i}\right)$$

$$\delta_x = 0.7 \text{ kJ}$$

Therefore, we can state the work done by the expanding gas $w = -5.5(7)$ kJ.

Note: This is a compound method of propagating errors; a single-step equation can be used, as explained in Toolkit A.8.

❓Exercise 3.8 Propagating Errors—Powers, Logarithms and Exponentials

---

Determine the value and corresponding uncertainty in the subject of each of the following equations:

1. $A = \log\dfrac{I_0}{I}$; $I = 0.602(5)$ and $I_0 = 1.000(5)$

2. $S = R\ln\dfrac{T_f}{T_i}$; $T_i = 353.7(5)$ K and $T_f = 378.1(5)$ K

3. $\dfrac{d[A]}{dt} = -k[A]^2$; $[A] = 0.475(9)$ M and $k = 3.48(5)$ mol^{-1} dm^3 s^{-1}

4. $[A] = [A]_0 e^{-kt}$; $[A]_0 = 0.25(2)$ M, $k = 0.0583(8)$ s^{-1} and $t = 30$ s (exactly)

5. $[H^+] = 10^{-pH}$; pH $= 4.258(6)$

6. $v = \sqrt{\dfrac{3RT}{M_r}}$; $M_r = 31.998$ g mol^{-1} and $T = 298(1)$ K

3.5 Identifying Outliers in a Single Data Series

Ordinarily, if a data point doesn't 'look correct', the most reliable test is to repeat the measurement to ensure that it is indeed a valid value for that data point. However, no matter how many repeat measurements we make, we can never be absolutely certain of a data point being an outlier; the Gaussian distribution will always have a statistical (finite) probability of a valid data point being far from the mean value.

A number of different techniques are used to identify outliers in data sets and those presented here are by no means exhaustive, each having specific applications in different settings. The common theme with each of these techniques is the use of *significance tables*, which are used to qualify what makes a point *statistically significant*.

We are showing these tests principally to indicate how such statistical verification of data may be used. It should be considered very rare for you ever to be thinking about 'removing' a data point; as mentioned earlier, it is preferable to repeat any measurement which seems to be erroneous.

> You can find such sources by loading your favourite internet search engine and searching for 'statistical tables'; however, be sure to use a reliable source for the data.

All three methods introduced here rely on tables of statistical data; we have included extracts of these statistical data here, but more exhaustive tables may be found in a range of sources, most of which are available from online resources.

3.5.1 Dixon's Test

Dixon's Q test, or sometimes just the 'Q test' is used for the identification and rejection of a *single* outlying data point in a set of data. It is one of the more common data verification methods used in analytical chemistry. It makes use of the *range* of the data (the difference between the maximum and minimum value data points) and the '*gap*', which is the difference between the questioned outlier and the next nearest data point:

$$Q = \frac{|\text{gap}|}{|\text{range}|} \tag{3.30}$$

> Remember that modulus signs '| |' around a number mean that you should ignore any negative signs and just consider the magnitude of the value.

This calculated value of Q is then compared with tables of *critical Q values* (Table 3.1), and if $Q > Q_{\text{crit}}$ the value may be rejected. When

Table 3.1 Dixon's Q_{crit} values used to eliminate outlying data with different levels of confidence[a].

Sample size	80%	90%	95%	99%
3	0.886	0.941	0.970	0.994
4	0.679	0.765	0.829	0.926
5	0.557	0.642	0.710	0.821
6	0.482	0.560	0.625	0.740
7	0.434	0.507	0.568	0.680
8	0.399	0.468	0.526	0.634
9	0.370	0.437	0.493	0.598
10	0.349	0.412	0.466	0.568
15	0.285	0.338	0.384	0.475
20	0.252	0.300	0.342	0.425
25	0.230	0.277	0.317	0.393
30	0.215	0.260	0.298	0.372

[a]A more complete version of this table with additional sample sizes and confidence intervals may be found in D. B. Rorabacher, *Anal. Chem.*, 1991, **63**(2), 139–146, DOI: 10.1021/ac00002a010.

reading the tables, it matters how many readings are in your data set and how confident you wish to be in knowing that your suspect value is (in all probability) a genuine 'outlier'. However, we can never be 100% confident of a data point being an outlier. We work through the application of Dixon's Q test in Example 3.10.

Example 3.10 Determining Whether a Value is an Outlier—Dixon's Q Test

Q: Seven students each measured the temperature change of a reaction. Can we be 95% confident that there is no outlier in the following data?

$$\Delta T / K: 2.92, 2.85, 2.91, 3.07, 2.92, 2.84, 2.94$$

A: To solve the problem, we need to undertake the following steps:

- Sort the data into ascending order and identify the potential 'outlier'.
- Calculate the 'range', 'gap' and Q for the data set.
- Compare the calculated value of Q with the corresponding value of Q_{crit} in the data table.

Firstly, let's sort the data into ascending order:

$\Delta T / K$: 2.84, 2.85, 2.91, 2.92, 2.92, 2.94, 3.07

All the data values appear to cluster around $\Delta T = 2.9$ K, leaving the suspect value $T = 3.07$ K.

Using eqn (3.30) to look at the 'gap' of the outlier from its nearest data point compared with the range of the data:

$$Q = \frac{|\,gap\,|}{|\,range\,|} = \frac{|\,3.07 - 2.94\,|}{|\,3.07 - 2.84\,|}$$

$$Q = 0.565$$

Having calculated a value of Q, we now read from Table 3.1 for seven values; to have a 95% confidence of a value being an outlier, Q must be greater than $Q_{crit} = 0.568$.

In this case Q (0.565) is not greater than Q_{crit} (0.568) if we want to be 95% confident, and so the value cannot be considered an outlier; therefore, it must remain in the data set and be used in calculations of mean, standard deviation and standard error.

Note: We could reject the data point if we only wanted to be 90% confident that it was an outlier, as in this case $Q_{crit} = 0.507$. This would allow us to reject the data point; however, we are less certain that it actually is an outlier!

Remember that the range should include the potential outlier.

It is tempting to apply this test to eliminate data points which look 'wrong'; however, eliminating data points makes our data set less robust, hence it is *always* better to go back and remeasure an apparently erroneous data point.

3.5.2 Grubbs' Test

Grubbs' test is also used to identify potential outliers in a data series, and works on the assumption that the data follows a Gaussian (normal) distribution (Figure 3.3). The Grubbs' test (eqn (3.31)) is a more robust mechanism of deciding on outliers than Dixon's Q test, but has a slightly more involved calculation:

$$G = \frac{|\,x_{out} - \bar{x}\,|}{s} \qquad (3.31)$$

This relates the suspected outlier (x_{out}) to the mean of all values ($\bar{x}$; this includes the suspected outlier) and the sample standard deviation of all values (s, (eqn (2.4)), again including the suspected outlier).

This calculated value of G is then compared with 'critical values' in data tables (Table 3.2). Again, if $G > G_{crit}$, we can discard the value as an outlier with the specified confidence. The process is shown in Example 3.11; you will notice that there are similarities with Dixon's Q test (Example 3.10).

Table 3.2 Grubbs' G_{crit} values used to eliminate outlying data with different levels of confidence.[a]

Sample size	90%	95%	97.5%	99%
3	1.148	1.153	1.155	1.155
4	1.425	1.463	1.481	1.492
5	1.602	1.672	1.715	1.749
6	1.729	1.822	1.887	1.944
7	1.828	1.938	2.020	2.097
8	1.909	2.032	2.126	2.221
9	1.977	2.110	2.215	2.323
10	2.036	2.176	2.290	2.410
15	2.247	2.409	2.549	2.705
20	2.385	2.557	2.709	2.884
25	2.486	2.663	2.822	3.009
30	2.563	2.745	2.908	3.103

[a]A more complete version of this table with additional sample sizes and confidence intervals may be found in F. E. Grubbs, *Technometrics*, 1969, **11**(1), 1–21, DOI: 10.2307/1266761 and F. E. Grubbs and G. Beck, *Technometrics*, 1972, **14**(4), 847–854, DOI: 10.1080/00401706.1972.10488981.

Remember to use the sample standard deviation because there are only a few measurements.

This is the same data set as used in Example 3.10.

Standard deviations take a long time to calculate by hand; use a spreadsheet or other package.

It is quicker and easier to calculate these values using a spreadsheet or in Python; it's good practice as it reduces the possibility of human error in the calculations.

Example 3.11 Determining Whether a Value is an Outlier—Grubbs' Test

Q: Seven students each measured the temperature change of a reaction. Determine the mean and standard error of the following data set, assuming that we are 95% confident about rejecting any potential outliers:

$$\Delta T / K: 2.92, 2.85, 2.91, 3.07, 2.92, 2.84, 2.94$$

A: As before, we break the problem into the following steps:

- Sort the data into ascending order and identify the potential 'outlier'.
- Calculate the mean and standard deviation of the full data set.
- Calculate G and compare the calculated value of G with the corresponding value of G_{crit} in the data table.
- Calculate the mean and standard deviation again, except with the outlier removed.

Let's now rank the data in ascending order:

$$\Delta T / K: 2.84, 2.85, 2.91, 2.92, 2.92, 2.94, 3.07$$

Here, $T = 3.07$ K is identified as a potential outlier. We then calculate the mean and standard deviation using the methods described in Chapter 2.

For this data set, $\bar{T} = 2.921$ K and $s = 0.076$ K; we now feed these values into eqn (3.31):

$$G = \frac{|\,x_{out} - \bar{x}\,|}{s} = \frac{|\,3.07 - 2.921\,|}{0.076}$$

$$G = 1.961$$

In this case, reading the 95% confidence value for seven data points gives $G_{crit} = 1.938$. Therefore, our calculated value of G is greater than G_{crit} and we can reject the data point $T = 3.07$ K.

As we have rejected a data point, we need to recalculate our mean and standard error:

$$\bar{T} = 2.90 \text{ K and } s = 0.0413 \text{ K}$$

Remember that to find the standard error we divide by the square root of the number of measurements; but now we only have six measurements, as we have eliminated one measurement as an outlier:

$$\sigma_{\bar{T}} = \frac{s}{\sqrt{N}} = \frac{0.0413}{\sqrt{6}}$$

$$\sigma_{\bar{T}} = 0.02 \text{ K}$$

After having excluded the outlier based on Grubbs' test, the average value for this data set is $\bar{T} = 2.90(2)$ K.

> Additional significant figures have been kept, as we are only at the midpoint of the calculation.

❓Exercise 3.9 Application of Dixon's Q Test and Grubbs' Test

For each of the following data sets, determine whether one of the values may be excluded as an outlier, firstly using Dixon's Q test and then using Grubbs' test at the specified confidence interval. Having decided whether to keep or reject a value, determine the mean and standard error in each data set.

1. A: 0.452, 0.445, 0.460, 0.420, 0.457; 95%
2. $\Delta H/\text{kJ mol}^{-1}$: −96.4, −99.2, −91.3, −104.9, −92.7, −94.0, −95.1; 90%
3. Titre$/\text{cm}^3$: 22.0, 23.1, 22.5, 22.0, 22.1, 22.5, 22.2; 95%
4. $\Lambda_0/\text{S cm}^2 \text{ mol}^{-1}$: 91.0, 92.3, 104.5; 99%
5. $\Delta T/\text{K}$: 2.55, 2.68, 2.71, 2.76, 2.73, 2.80, 2.79, 2.72; 90%
6. $\varepsilon/10^3 \text{ M}^{-1} \text{ cm}^{-1}$: 102.4, 95.2, 77.7, 104.5, 90.0, 88.3, 105.6, 94.3; 80%

3.6 Comparing Data Series—Student's *t* Test

Often in science, we want to know whether an 'intervention' has had an effect or not. From knowing if it is worth fertilising crops, to determining if drug treatments are effective or even if a particular educational approach improves student learning, Student's *t* test[††] is well established in biology and pharmacology and it is increasingly used in chemistry by those working at the interface of the discipline. Many spreadsheet packages and data analysis resources will automatically calculate values for Student's *t* test, but the process involved in the calculation (and the meaning of the results once obtained) are explained here.

The test examines two data sets (usually two independent tests, such as the rates of reaction with two different catalysts) and determines whether any differences between the data sets are statistically significant (for instance, testing whether a catalyst actually works). The *t* test also has a variation called a 'paired' test, where we can re-examine the same data set under a different set of conditions. An example would be comparing the activity of a new catalyst with its activity after use for a number of reactions; the paired *t* test would demonstrate any statistically significant differences between the data sets. The problem in each case is largely the same and the same process can be used to determine the statistical significance of any difference between a pair of data sets.

We have said previously that increasing the number of measurements made reduces the standard error and we obtain a distribution approaching the normal distribution; for such large sets of data, the chance of any one measurement being outside the range 'mean ± two standard deviations' ($\bar{x} \pm 2\ s$) is less than one in twenty (see Table 2.3 for these limits based on standard deviations). However, it is extremely unusual to take *every* measurement of a population; instead, we undertake a process called *sampling*, in which just a few measurements are actually made relative to all possible measurements which *could* be made. Therefore, the calculation of both the mean and the standard error is just an estimate based on a few measurements of the variable.

A normal or Gaussian distribution has the classic bell-shaped distribution of data (Figure 3.4).

Student's *t* test uses the idea that since we only ever *sample* a small proportion of all possible information, we have to draw our conclusions with a level of confidence based on a limited sample size. The test relies on the assumption that data are distributed normally; the test then examines whether any differences in the two data series are due to random chance in sampling (Figure 3.4), or if there are in fact genuine differences between the two sets of data.

[††]This test was developed by W.S. Gosset who worked in the Guinness brewery, but he published it under the pseudonym 'Student'—hence its name!

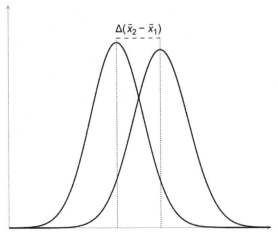

$\Delta(\bar{x}_2 - \bar{x}_1)$

Figure 3.4 Two data sets of sampled data: Student's t test determines whether the difference in the means of the two data sets is significant, or could just be due to random error.

Student's t test (eqn (3.33)) examines two data sets and compares the 'difference between the means' and the 'variability in the data within a group'. When t is calculated, it is again compared with data tables. If our calculated value of t is larger than the corresponding value in the data table, this means that any difference between the data sets *is* statistically significant, and we can say that we *'reject the null hypothesis'*.[††]

When reading data from Table 3.3, we need to be familiar with a few terms. Firstly, the number of 'degrees of freedom' is linked to how many samples (from both populations) have gone into the test:

$$\text{degrees of freedom} = (n_1 + n_2) - 2 \qquad (3.32)$$

We then need to know if we are making a 'one-tailed' or 'two-tailed' comparison, that is, if the difference occurs only on one side of the distribution ('one-tailed') or on both sides of the distribution ('two-tailed'). In the case of the two-tailed comparison, the probability of statistical differences between the two data sets is spread over both extremes of the distribution. The table of t test confidence values (Table 3.3) works in much the same way as for the Dixon's Q and Grubbs' tables, but now giving an indication of a statistically relevant difference between the data sets :

$$t = \frac{|\bar{x}_1 - \bar{x}_2|}{\sqrt{\dfrac{s_1^2}{n_1} + \dfrac{s_2^2}{n_2}}} \qquad (3.33)$$

[††]The 'null hypothesis' in a statistical test simply says that there is no statistically significant difference between two populations of data.

Table 3.3 Student's t table showing both confidence percentage and p-value thresholds for one- and two-tail versions of the test.[a,b] The degrees of freedom may be found by d.o.f $= (n_1 + n_2) - 2$.

Degrees of freedom	p (one tail)			
	90% / 0.10	95% / 0.05	97.5% / 0.025	99.5% / 0.005
	p (two tails)			
	80% / 0.20	90% / 0.10	95% / 0.05	99% / 0.01
1	3.078	6.314	12.71	63.66
2	1.886	2.920	4.303	9.925
3	1.638	2.353	3.182	5.841
4	1.533	2.132	2.776	4.604
5	1.476	2.015	2.571	4.032
6	1.440	1.943	2.447	3.707
7	1.415	1.895	2.365	3.499
8	1.397	1.860	2.306	3.355
9	1.383	1.833	2.262	3.250
10	1.372	1.812	2.228	3.169
15	1.341	1.753	2.131	2.947
20	1.325	1.725	2.086	2.845
25	1.316	1.708	2.060	2.787
30	1.310	1.697	2.042	2.750
40	1.303	1.684	2.021	2.704

[a]A more complete version of this table with additional sample sizes and confidence intervals may be found in Table III of R. A. Fisher and F. Yates, *Statistical Tables for Biological, Agricultural and Medical Research*, Oliver & Boyd, Edinburgh/London, 6th edn, 1963.

[b]The percentage indicates the probability of the null hypothesis (*i.e.* that that there is no difference between the data sets), while the p value, reported by most spreadsheets, indicates a probability of a difference between the sets.

While this equation looks complex, in fact, we have seen some of this before. The term s^2/n is the square of the standard error for the data set (see eqn (3.9)), so we can rewrite eqn (3.33) in the form shown in eqn (3.34) instead, using the standard error for each data series:[§§]

$$t = \frac{|\bar{x}_1 - \bar{x}_2|}{\sqrt{(\sigma_{\bar{x}_1})^2 + (\sigma_{\bar{x}_2})^2}}$$

(3.34)

As with the examples for Dixon's Q and Grubbs' tests, we demonstrate the application of Student's t test in Example 3.12, since it is easier to understand the process by seeing it in action.

[§§]We include the more complex representation because this is how it is represented in most texts.

Example 3.12 The Effectiveness of a Catalyst

A number of students each use the method of initial rates to determine the rate of a reaction with one of two catalysts. Their data are combined to give a data set for each catalyst, where the mean and standard deviation can be determined for each set of data.

To determine whether there are any statistical differences between the two catalysts (catalyst A and catalyst B), Student's t test is performed. The data collected by the class are as follows.

	Catalyst A rate / mol dm^3 s^{-1}	Catalyst B rate / mol dm^3 s^{-1}
	0.520	0.594
	0.594	0.631
	0.415	0.418
	0.633	0.552
	0.444	0.654
	0.618	0.524
	0.456	0.628
	0.609	0.640
	0.681	0.582
	0.550	0.498
	0.521	0.599
	0.563	0.582
	0.587	0.537
	0.486	0.628
	0.435	0.610
	0.661	0.591
	0.620	0.605
	0.548	0.580
	0.607	0.613
	0.412	0.596
	0.489	0.620
$\bar{x}$	0.543	0.585
s	0.081	0.053

We use these data to determine t for each data series:

$$t = \frac{|\bar{x}_1 - \bar{x}_2|}{\sqrt{\dfrac{s_1^2}{n_1} + \dfrac{s_2^2}{n_2}}} = \frac{|0.543 - 0.585|}{\sqrt{\dfrac{1}{21}(0.081^2 + 0.053^2)}}$$

$$t = 1.79$$

We have 21 measurements for catalyst A and 21 for catalyst B. We use this to calculate the degrees of freedom of the data series:

$$\text{d.o.f.} = n_A + n_B - 2$$
$$= 21 + 21 - 2$$
$$= 40$$

Having calculated the number of degrees of freedom, we now read from Table 3.3; for a two-tailed distribution, with $p = 0.05$ (or 95% confidence that there is no statistical difference between the data sets), the critical value is 2.021. Since our value of t is less than this, there is not enough evidence to suggest that the differences between the two data sets are significantly relevant, and we conclude (with 95% confidence) that they are most probably two samples from the same overall population.

Applying these statistics to the chemistry of the situation, the data indicate that both catalyst A and catalyst B have the same activity. We could also go further to look at catalyst degradation by using a paired t test of data from two points in a catalyst's life.

💬
The value of p calculated using the 'TTEST' function on a spreadsheet is $p = 0.0733$; this would indicate that the differences would not fulfil the 'no difference —null' hypothesis 8% of the time (rounding up).

3.7 Summary

Chemistry is a practical subject; when we report our results, we need to ensure that we are reporting our values appropriately. An experimentally determined value given with correct and appropriate precision communicates more than just the value itself; it tells us about the nature of the experiment in which it was determined, as well as how effective the value is likely to be when we go on to use it in other experimental models.

Fundamentally, error analysis has its roots in statistics and, with enough measurements, random errors will—eventually—average to zero. However, it is rarely practical to make enough measurements to cover an entire population, necessitating the use of statistical models based on our 'sampling' approach.

Summary : Chapter 3

- A precise measurement is repeatable; an accurate measurement is close to the 'true' value.
- High precision and low accuracy indicates a significant systematic error in the experiment; low precision and high accuracy indicates significant random error in the experiment.
- You cannot read a piece of scientific equipment to a higher precision than its scale.
- If a scale reading is midway between graduations, such that it cannot be rounded with any certainty, you should round to the even valued graduation to prevent bias.

Application of Significant Figures

- To avoid confusion over how many significant figures are reported, you should use standard form (scientific notation) or SI prefixes with units.
- When adding or subtracting values, preserve the number of decimal places of the least precise initial value.
- When multiplying or dividing values, preserve the number of significant figures of the least precise initial value.
- When finding the logarithm of a value you gain a significant figure and when finding an exponent you lose a significant figure, unless the modulus of the value is less than the base; then you preserve significant figures.

Errors and Uncertainties

- The standard error is found from the standard deviation, is reported to a single significant figure (rounded up!) and becomes smaller with a greater number of measurements.
- The uncertainty in the final significant figure is indicated in brackets; e.g. $T = 34.28(7)$ K is equivalent to $T = 34.28 \pm 0.07$ K.
- Uncertainties in measured quantities combine (propagate) differently, depending on the mathematical relationship between the quantities.
- Data points which are perceived to be outliers should be re-measured if at all possible, and not simply disregarded.
- Perceived outliers should be tested for statistical significance using an appropriate statistical test and should only be removed from the data set if they fail that test.

Key Equations

In all error propagations, x, y, z are measurements, a is the calculated quantity, δ denotes the uncertainty in a particular value and b is a constant base (logarithms and exponentials).

- The standard error:

$$\sigma_{\bar{x}} = \frac{s}{\sqrt{N}}$$

- Propagating errors: addition

$$a = x + y - z \cdots$$
$$\delta a \approx \delta x + \delta y + \delta z \cdots$$

- Propagating errors: multiplication or division

$$a = \frac{xy}{z}$$

$$\frac{\delta a}{|a|} \approx \frac{\delta x}{|x|} + \frac{\delta y}{|y|} + \frac{\delta z}{|z|} \cdots$$

- Propagating errors: powers

$$a = x^n$$

$$\frac{\delta a}{|a|} = \left| n \frac{\delta x}{x} \right|$$

- Propagating errors: logarithms

$$a = \log_b x$$

$$\delta a = \frac{\delta x}{\ln(b) x}$$

- Propagating errors: exponentials

$$a = b^x$$
$$\delta a = \ln(b) \times a \delta x$$

- Dixon's Q value:

$$Q = \frac{|\text{gap}|}{|\text{range}|}$$

- Grubbs' G value:

$$G = \frac{\left| x_{out} - \bar{x} \right|}{s}$$

- Student's t value:

$$t = \frac{\left| \bar{x}_1 - \bar{x}_2 \right|}{\sqrt{\dfrac{s_1^2}{n_1} + \dfrac{s_2^2}{n_2}}}$$

❓Exercise 3.10 Experimental Uncertainty in a Chemical Context

---

1. Five students each measured the limiting molar conductivity, Λ^0, of HCl:

$\Lambda^0 / \text{S cm}^2 \text{ mol}^{-1}$: 426.8, 432.8, 421.26, 371, 415.1

Using Dixon's Q test, determine whether there are any outliers, to 95% confidence, and an appropriate mean value.

2. Determine the mean and standard error of the following data set:

   $\Delta H / \text{kJ mol}^{-1}$: −98.5, −91.4, −102.7, −99.9, −93.6, −102.3, −104.9

3. The rotational constant, B, for $^{12}C^{16}O$ is measured as 1.923(5) cm^{-1}. Using equation (3.35) determine the bond length, r, and its associated error.

$$B = \frac{h^2}{8\pi\mu r^2} \qquad (3.35)$$

   where:

$$\mu = \frac{m_1 m_2}{m_1 + m_2}$$

> Remember to check the units; you will need to convert the wavenumber to energy units.

4. The temperature change of a reaction when different quantities of solid barium hydroxide octahydrate was added to an excess of ammonium chloride was recorded.

m/g	ΔT/K
47.24	14.7
47.38	14.9
47.28	14.2
47.15	13.4
47.45	15.0

Using Grubbs' test, determine whether there are any outliers to 95% confidence, and consequently determine the molar enthalpy of reaction and its associated error.

$$c = 24.85(8) \text{ J K}^{-1} \text{ mol}^{-1}$$

> You will need to determine $\Delta T / m$ before you work out whether there are any outliers, because the masses are different in each case.

5. A circular 'coin' of silver was measured to be 31.45 mm in diameter using vernier callipers with 20 μm graduations. The coin was determined to be 3.657 mm in depth using a micrometer with 2 μm graduations. Given that the density of silver is 10.54(6) × 10^3 kg m^{-3}, what is the expected mass and uncertainty in the mass of the coin?

Solutions to Exercises

Solutions: Exercise 3.1

1. 2
2. 3
3. 5
4. 4

5. 11
6. 8
7. 5
8. ∞

Remember the rule, 'round to even'.

The unit mol^{-1} dm^3 s^{-1} is the unit of a second-order rate constant

Solutions: Exercise 3.2

1. $T = 298.2$ K
2. $M_r = 40.00$ g mol^{-1}
3. $E_{cell} = 1.095$ V
4. $\Lambda_{NaOH} = 24.8$ S cm^2 mol^{-1}

5. $\Delta_c H_{diamond} = -395.40$ kJ mol^{-1}
6. $\Delta C_{p,m} = -30.86$ J K^{-1} mol^{-1}
7. $\Delta H = -37.68$ kJ mol^{-1}
8. $p = 103\,067$ Pa

Solutions: Exercise 3.3

1. $q = 3.3$ kJ

2. $K = 2.5$
3. $m = 1.15$ kg
4. $p = 2.09$ MPa or 20.9 bar

5. $k_d = 7.424 \times 10^6$ mol^{-1} m^3 s^{-1} or 7.424×10^9 mol^{-1} dm^3 s^{-1}
6. $c = 0.2486$ mol dm^{-3}
7. $m = 39.4$ g
8. $n = 15.0$ mmol

Solutions: Exercise 3.4

1. NaCl: $M_r = 58.44$ g mol^{-1}, 0.024 8 mol
2. CH_3CH_2OH: $M_r = 46.068$ g mol^{-1}, 0.050918 mol
3. $HOOCCH_2CH_2COOH$: $M_r = 118.087$ g mol^{-1}, 0.027 3 mol
4. C_6H_5COONa: $M_r = 144.104$ g mol^{-1}, 407 μmol
5. C_8H_6S: $M_r = 134.20$ g mol^{-1}, 0.077 50 mol
6. $C_{61}H_{81}ClN_3S$: $M_r = 319.85$ g mol^{-1}, 12 μmol
7. $C_{82}H_{13}N_2O_3Cl$: $M_r = 479.01$ g mol^{-1}, 174 μmol
8. $[Ru(C_{10}H_8N_2)_3]^{2+}$: $M_r = 569.63$ g mol^{-1}, 14 μmol

Solutions: Exercise 3.5

1. $A = 0.077$ 3(3)
2. $p = 140(3)$ mm Hg
3. $I = 1.45(2)$ mA
4. $V = 2.82$ V

5. $\Delta T = 2.37(5)$ K
6. $m = 1.000(1)$ g
7. $T = 32.2(2)$ °C
8. $\Delta H = -108(3)$ kJ mol^{-1}

Solutions: Exercise 3.6

1. $\Delta T = -4.0(3)$ °C

2. $m = 0.225\ 9(2)$ g

3. $\Delta_r H = 113(4)$ kJ mol^{-1}

4. $\Delta C_p = 63.5(5)$ J K^{-1} mol^{-1}¶¶

Solutions: Exercise 3.7

1. $p = 5.024(8)$ MPa

2. $V = 3.25(4) \times 10^{-3}$ m^3

3. $p = 101(1) \times 10^3$ Pa

4. $E = 486(8) \times 10^{-19}$ J

5. $C = 0.39(2)$ kJ K^{-1}

6. $[NaCl] = 0.877\ 7(4)$ mol dm^{-3}

7. $n = 0.533(4)$ mol

8. $S = 22(1)$ J K^{-1} mol^{-1}

Solutions: Exercise 3.8

1. $A = 0.22(2)$

2. $S = 0.55(2)$ J K^{-1} mol^{-1}

3. $\dfrac{d[A]}{dt} = -0.8(2)$ mol dm^3 s^{-1}

4. $[A] = 0.04(2)$ M

5. $[H^+] = 55.2(7)$ μM

6. $v = 482(1)$ m s^{-1}

Solutions: Exercise 3.9

1. Dixon's: keep; $A = 0.447(8)$

   Grubbs': remove (just!); $A = 0.454(4)$

2. Dixon's: keep; $\Delta H = -96(2)$ kJ mol^{-1};

   Grubbs': remove $\Delta H = -95(2)$ kJ mol^{-1}

3. Dixon's: keep

   Grubbs': keep; titre = 22.3(2) cm^3

4. Dixon's: keep

   Grubbs': keep; $\Lambda_0 = 96(5)$ S cm^2 mol^{-1}

5. Dixon's: remove

   Grubbs': remove; $\Delta T = 2.74(2)$ K

6. Dixon's: keep

   Grubbs': keep; $\varepsilon = 94(4) \times 10^3$ M^{-1} cm^{-1}

¶¶When adding and subtracting terms, sometimes we can ignore uncertainties which are very small (as they are negligible compared with other terms).

Solutions: Exercise 3.10

1. 371 S cm^2 mol^{-1} removed; $\Lambda^0 = 420.0$ S cm^2 mol^{-1}
2. $\Delta H = 99(2)$ kJ mol^{-1}
3. $r = 1.131(3)$ Å
4. No outliers; $\Delta H = +140(4)$ kJ mol^{-1}
5. $m = 29.9(2)$ mg

Learning Points: What We'll Cover

- Recording data in tables in an unambiguous manner and why table headers are the way they are
- How to present a graph effectively, what variable to plot on which axis and how to represent units
- Using linear relationships to gain meaning from graphs
- Rearranging non-linear equations into a linear form and plotting manipulated variables to gain a linear plot
- How to determine the gradient, intercept and the uncertainty of a straight-line plot
- Determining a molecular property using a straight-line graph

Why This Chapter Is Important

- Plotting a graph is a key experimental skill; whether plotting by hand or on a computer, the same principles must be followed.
- Many relationships in chemistry are inherently non-linear; by reworking the equation to make it a 'linear form', we make it easier to identify trends in our data when the appropriately transformed experimental data are plotted on a graph.
- A single data point cannot test a theory – plotting a graph allows for the 'averaging out' of experimental 'noise' (random data fluctuations), and is essential for fitting experimental data to models.

Tables and Graphs— Presenting and Analysing Experimental Data

Presentation of data in a way that other people can understand is fundamental. A set of conventions published by IUPAC (the International Union of Pure and Applied Chemistry)—the governing body of chemistry—unifies the way that molecules are named, the names of units and how data should appear in tables and graphs. The whole reason for these conventions is that we should all be able to understand the data presented to us so that there is no ambiguity.

In this chapter, we will introduce you to these IUPAC conventions (which may be contrary to your current understanding), and explain why they actually make a huge amount of sense once you understand how they work!

Knowing what we need to plot in order to find specific values allows us to test our theories in the laboratory. Much chemical data comes from straight-line plots of the form '$y = mx + c$', so being able to manipulate data to fit this form allows us to readily see a trend in our data, making it a very powerful technique for data analysis.

4.1 Recording Data

When we are in the lab, particularly when we are doing physical or analytical chemistry experiments, we have to record a lot of data relatively quickly. Usually, we start with the best intentions, clearly labelling the columns, perhaps even drawing the table with a ruler. Our first entries in the table in our lab book are meticulously neat but, for many, these neat columns of data become increasingly scruffy; gone are the ruled lines, the handwriting becomes less careful and, if a new page is started, the column headings often don't get repeated.

In recording our data, we must be careful not to accidentally create ambiguity—particularly with regard to units or scaling factors. The raw record of data in our laboratory notebooks must be represented consistently and unambiguously using the appropriate IUPAC rules on data presentation. These rules give a common understanding across a wealth of literature to minimise ambiguity.

4.1.1 Presenting Data in Tables

The IUPAC rules for presenting data in tables in order to minimise ambiguity are:

1. All columns should have clearly labelled headings, with units placed alongside (not next to the data in the table).
2. If there are multipliers for values (such as $\times 10^{-6}$), these should be grouped in the header as far as possible. Sometimes, when data runs across a number of orders of magnitude, this isn't possible, but try put as much information in the header as is feasible.
3. Units and multipliers appear with a 'stroke' (a 'division', formally a *solidus*, /), not in brackets.

This last point is usually the one that is least familiar to chemists, but it makes perfect sense if you think of a table as shorthand for a lot of equations.

The actual formatting of the table is largely down to preference, provided you follow these rules. Table 4.1 has been presented with a double line between the header and the contents; while some tables use this convention, it is simply a personal preference. However, laying out the table with this style can demonstrate why the solidus (/) is used rather than a bracket.

If you think of the table as containing a number of equations, then this 'double line' can be thought of as an equals sign. For example, the measurement

$$\frac{T}{°C} = 9.1 \tag{4.1}$$

rearranges as

$$T = 9.1°C$$

Table 4.1 Vapour pressure of ethyl acetate at different temperatures. Data derived from *Lange's Handbook of Chemistry*, 15th edn and *CRC Handbook of Chemistry and Physics*, 100th edn.

T/ °C	p / mm Hg
9.1	40.0
27.0	100
77.1	760
⋮	⋮

Table 4.2 Arrhenius data for the decomposition of HI.

T/K	k/10⁻⁶ dm³ mol⁻¹ s⁻¹	1/T/10⁻³ K⁻¹	ln k
556	0.704	1.80	−14.17
575	2.44	1.74	−12.92
629	60.4	1.59	−9.715
⋮	⋮	⋮	⋮

Each number in the table is just that: a unitless value, and it is only by combining it into an equation that it gets its unit and makes sense as a value. Thinking about tables in this way allows the more unusual column headers to make sense, as in Table 4.2.

Here we see a number of more complicated column headers; these start to make sense when we look at the lone values in the table as parts of equations. We can treat the first column (temperature) in the same way as we did in eqn (4.1). In the second column, however, we have a scaling factor *and* a compound unit:

$$\frac{k}{10^{-6}\,\mathrm{dm^3\,mol^{-1}\,s^{-1}}} = 0.704 \qquad (4.2)$$

This rearranges in exactly the same way to give the rate constant k:

$$k = 0.704 \times 10^{-6}\,\mathrm{dm^3\,mol^{-1}\,s^{-1}}$$

In the case of the third column, $1/T/10^{-3}\,\mathrm{K^{-1}}$:

$$\frac{1/T}{10^{-3}\,\mathrm{K^{-1}}} = 1.80 \qquad (4.3)$$

the column header would be equally valid if written as $10^3\,\mathrm{K}/T$; either way this rearranges to become:

$$1/T = 1.80 \times 10^{-3}\,\mathrm{K^{-1}}$$

You will have noted that the column with ln k has no unit; the reason for this is explained in Chapter 1.

❓Exercise 4.1 Identifying Data from Tables

For the following values, you can interpret them as having been 'read' from a table. Rearrange each equation to express the variable as its value and unit.

1. $[\mathrm{HCl}]/\mathrm{mol\,dm^{-3}} = 0.150$

2. $10^{-5}p/\mathrm{Pa} = 8.76$

3. $E/\mathrm{mV} = 316$

4. $10^{-4}\,K = 4.89$

5. $\tau/\mathrm{ns} = 5.7$

6. $k/10^{-3}\,\mathrm{dm^3\,mol^{-1}\,s^{-1}} = 5.2$

7. $\Delta H/\mathrm{kJ\,mol^{-1}} = -458.2$

8. $[\mathrm{A}]/\mathrm{mol\,dm^{-3}} = 0.040$

4.1.2 Presenting Data on Graphs

The same rules about presenting units on tables are also used on graphs. In short, units should appear behind the 'stroke' again (solidus, Figure 4.1). Again, the reason is that we can only plot a 'value' on the graph (rather than data with its unit) and therefore this value needs to have context added to its description. Additionally, there is no reason why any graph needs to have an axis starting at zero. Indeed, for many cases in chemistry, the values we record go nowhere near zero.

We can now move on to use data to determine useful values in chemistry.

4.2 Straight-line Graphs: '$y = mx + c$'

As chemists, we usually draw a graph in order to use our recorded data to determine another value, using either the gradient or the intercept. The challenge is, 'What do we plot on the graph to get a straight line?'

To get useful values of the gradient or intercept, we usually need to have a straight-line graph of the form '$y = mx + c$', where m and c are the gradient and intercept, respectively. However, we need to decide what to plot on the x- and y-axes.

4.2.1 Putting Quantities on the Correct Axes

There is a straightforward method for assigning our measurements to the appropriate axes; when conducting an experiment, there is always a variable we have control over and another variable which is measured.

1. The variable you have control over (or something related to it, see Section 4.2.3) should be plotted on the x (or horizontal) axis. The fancy term for this is the *independent* variable.

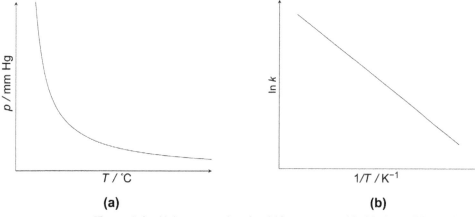

(a)   **(b)**

Figure 4.1   Units on graphs should be presented behind a solidus, '/'.

2. The 'thing you measure' (or something related to it, see Section 4.2.3) should be plotted on the y (or vertical) axis. The fancy term for this is the *dependent* variable.

3. Constants should not usually be included on values of either axis. (In Example 4.1, we plot $\ln k$ against $1/T$, *not* $1/RT$. Instead the gas constant must be considered when we look at the gradient.)

For example, when conducting a calorimetry experiment in the lab, we might record the temperature every 30 s for 5 min before adding a reagent, and continue to record the temperature every 30 s for a further 10 min after the reagent is added. When we draw a graph, we identify the variable over which we have control (the time) and record this on the x-axis, while the variable we measure (the temperature) goes on the y-axis (Figure 4.2).

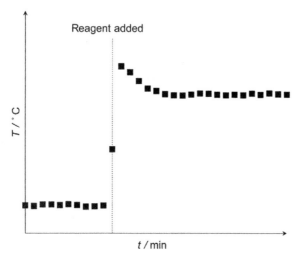

Figure 4.2   In this experiment, the temperature is measured at fixed intervals; this makes the time the independent variable, plotted on the x-axis, while the temperature is the dependent variable, plotted on the y-axis.

This graph doesn't follow the pattern of a '$y = mx + c$' graph, because it is mainly plotted to make the data clear and easier to analyse, however the rules as to which variable is plotted on which axis are the same.

4.2.2   Graphs of the Form '$y = mx + c$'

In chemistry, we normally plot graphs based on fitting the data to mathematical relationships. However, very few relationships in chemistry naturally come arranged in the form '$y = mx + c$', and

Remember that the notation $[A]_0$ is the initial concentration of the reactant A, *i.e.* the concentration you put in to the reaction.

The constant, c, can be positive, negative or even zero.

Since a rate constant, k, is always a positive value, when an equation appears with $-kt$ you know that a sketch of the graph will have a negative (downwards) gradient. Another way of thinking about this is that when we see a negative gradient for the term $-kt$ the two negatives cancel out, giving a positive value of k.

typically we have to do a little bit of rearranging in order to get an equation of this form.

The integrated rate equation for zero-order kinetics, however, is an equation which naturally fits the 'straight-line graph form'. Looking at eqn (4.4), it appears fairly straightforward:

$$[A] = [A]_0 - kt \qquad (4.4)$$

Let's think first about what we would control during an experiment. It should hopefully be clear that we are again in control of the time measurement, and we are in some way measuring the concentration of reagent remaining after that time. To use the fancy terminology introduced earlier, the time t is the *independent* variable, while the concentration of the reagent $[A]$ is the *dependent* variable.

Let's look at the equation in more detail and it becomes clear how the '$y = mx + c$' appears:

$$
\begin{array}{ccccc}
y & = & c & + & m & x \\
\downarrow & & \downarrow & & \downarrow & \downarrow \\
[A] & = & [A]_0 & - & k & t
\end{array}
\qquad (4.5)
$$

Transferring this to a graph (and following the form of the '$y = mx + c$' equation) means that if we plot the time on the x-axis and the concentration of the reagent $[A]$ on the y-axis, the gradient from this graph will be $-k$ (note the minus sign in the equation) and the intercept is the initial concentration of A, the concentration of A when $t = 0$ (Figure 4.3).

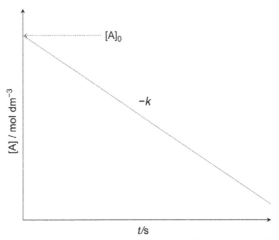

Figure 4.3   Following the form of the equation for zero-order kinetics, the gradient of the graph is $-k$ and the intercept is $[A]_0$.

Another example of this direct straight-line relationship is the expression of Charles' law (one of the laws which combine to give the ideal gas law). Charles observed that the volume of a gas was directly proportional to its temperature ($V \propto T$). If we state this in terms of what Charles changed and measured, we can have an idea of what to plot; an increase in temperature will lead to an increase in volume (at constant pressure). Knowing the ideal gas law, we can now rearrange the ideal gas law to the form where V is the dependent variable (eqn (4.6)).

Again, we can analyse the equation to see the '$y = mx + c$' form (eqn (4.6)); however, this time we see that there is no $+c$ term in the theory, so we would expect a perfect line of best fit to go through (0, 0) (Figure 4.4)

$$y \;=\; m \quad x \;+\; c$$
$$\downarrow \qquad \downarrow \quad \downarrow$$
$$V \;=\; \frac{nR}{p} \; T \tag{4.6}$$

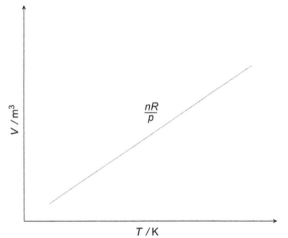

Figure 4.4   Following the form of the equation for Charles' law, we see the predicted intercept of 0 and that the gradient is $^{nR}/_{p}$.

4.2.3   Graphs of Non-linear Functions—Rearranging Equations to Fit the Form '$y = mx + c$'

As previously mentioned, many mathematical relationships in chemistry are not as simple as our independent and dependent variables being directly proportional to one another. Instead, we need to manipulate our equation to fit it into the form of a straight-line graph '$y = mx + c$'. The same rules should be followed as outlined in Section 4.2.2; the case of the Arrhenius equation is shown in Example 4.1.

Note that you should *never force a graph to go through zero*, even if you have measured this as a data point; when plotting a line of best fit, any measured data point at 'zero' should be treated the same as any other. Just because the theory predicts an intercept of zero doesn't mean the measurement at zero has any greater significance.

Example 4.1 Plotting the Arrhenius Equation as a Straight-line Graph

You are probably already familiar with the Arrhenius equation, relating the rate constant for a reaction, k, to the absolute temperature, T. The Arrhenius equation itself is:

$$k = Ae^{-E_a/RT} \qquad (4.7)$$

where E_a is the activation energy for the reaction, A is the 'pre-exponential factor' (an empirical constant) and R is the gas constant.

If data for this are plotted, the graph will look curved (an exponential decay), Here we are plotting the temperature T on the x-axis, as we can control the temperature, and the rate constant for the reaction k is plotted on the y-axis.

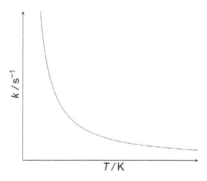

If this equation is rearranged, we are still looking to plot T (or something related to it) on the x-axis. Thinking back to Chapter 1, to eliminate an exponential, we need to take a natural log (ln), so we rearrange eqn (4.7) to isolate the exponent term and then take logs of both sides:

$$\frac{k}{A} = e^{-E_a/RT}$$

$$\ln\frac{k}{A} = \frac{-E_a}{RT}$$

This is not yet in a form that we can plot, because it is unlikely that we have a value for A, but we can use the 'rules of logs' (Chapter 1) to convert this to the traditional $y = mx + c$ form:

$$\ln k - \ln A = \frac{-E_a}{RT}$$

Following a quick rearrangement, we can see how the $y = mx + c$ appears:

$$
\begin{array}{ccccc}
y & = & c & + & m & x \\
\downarrow & & \downarrow & & \downarrow & \downarrow \\
\ln k & = & \ln A & - & \dfrac{E_a}{R} & \dfrac{1}{T}
\end{array}
\qquad (4.8)
$$

Thus, a straight-line plot will result if $^1/_T$ (in kelvin!) is plotted on the x-axis and $\ln k$ is plotted on the y-axis. This will give an intercept of $\ln A$ and a gradient of $-E_a/_R$.

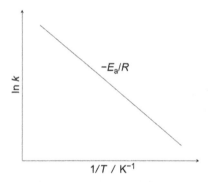

The linearised form of the Arrhenius plot can easily be used to determine the activation energy for the reaction, E_a.

Normally it is difficult to get the intercept directly from this plot, as the value $x = 0$ is usually a long way off the scale from where values of $^1/_T$ are plotted.

If you've plotted $^1/_T$ on the x-axis and your graph still looks curved, it may be that you've forgotten to convert temperatures to kelvin.

? Exercise 4.2 Identifying Appropriate Plots

Sketch linear '$y = mx + c$' graphs for the following equations, using the correct variables on each axis, including appropriate units. On your sketch, clearly indicate the gradient and intercept in each case.

1. Show how absorption, A, changes with concentration, c:

$$A = \varepsilon cl$$

A has no unit.

2. Show how pressure, p, changes with temperature, T:

$$pV = nRT$$

3. Show how pressure, p, changes with volume, V:

$$pV = nRT$$

4. Show how concentration, [A], varies with time, t, in a zero-order process:

$$[A] = [A]_0 - kt$$

5. Show how concentration, [A], varies with time, t, in a first-order process:

$$[A] = [A]_0\, e^{-kt}$$

6. Show how concentration, [A], varies with time, t, in a second-order process:

$$\frac{1}{[A]} = \frac{1}{[A]_0} + kt$$

7. Show how fluorescence lifetime, τ, in nanoseconds, varies with concentration of quencher, [Q]:

$$\frac{1}{\tau} = \frac{1}{\tau_0} + k_q [Q]$$

8. Show how vapour pressure, p, changes with temperature, T:

$$\ln p = \frac{-\Delta_{vap} H}{RT} + C$$

9. Show how equilibrium constant, K, changes with temperature, T:

$$\Delta H - T\Delta S = -RT \ln K$$

10. Show how cell potential, E, changes with temperature, T:

$$\Delta H - T\Delta S = -nFE$$

Cell potential has the unit V (or J C^{-1}).

4.3 Determining Gradients and Intercepts

As you have seen from the examples, graphs allow us to derive new values from either the gradient of the line of best fit (also referred to as the slope) or the intercept of this line with the y-axis. Any value of the intercept should be taken when the x-value is zero, and it has the same units as is on the y-axis. When drawing a graph by hand, it may not be feasible for this value to appear on the plot, and so in this section we will show you how this can be found (a process called *extrapolation*).

The gradient should be determined by 'fitting' the line of best fit through the data. It should be rare that any data should be discarded from this set (see Chapter 3 for guidance on statistical modelling). While it is all too easy to claim a given point as an 'outlier' and ignore it, it is not very scientific to do so.[†] As a minimum, though, you need five data points to draw a line of best fit with any level of confidence, and you should aim to record more than this.

Real data contain *noise*. Figure 4.5 shows the random fluctuations which exist no matter how carefully we do our experiments. The presence of noise in our measurements makes it essential that we plot a graph to see the effect on our measurements. When we plot the line of best fit, we are effectively 'averaging out' the random noise and should be left with a gradient which accurately represents the data trend in the absence of the noise. The more data points we are able to record, the more effectively the noise is 'averaged out' by the line of best fit and the more reliable our result.

The line of best fit should balance the distance that points are above, below and to each side of it (Figure 4.6). Practically, you only need to worry about the spread of the data either above and below (vertical, or *y* displacement) *or* to each side (horizontal, or *x* displacement), as they are both doing the same thing.

Although most graphs are now drawn on computers, it is useful to know what the computer is doing to fully understand the results. It is also good to know for when computers are unavailable (for instance, if plotting a graph under examination conditions!).

At no point should we ever attempt to model our data based on a single measurement or even the average of a limited selection of our data. Without the full data set, we cannot tell whether the noise has made these data points 'higher' or 'lower' than the model would predict. All data *must* be seen in context.

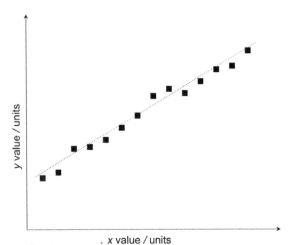

Figure 4.5   Real data contain noise. These fluctuations are normal, but mean that data points themselves should not be used directly to try to calculate further values.

[†]If in doubt when analysing experimental data it is always best to go back and remeasure that point, just in case an error was made. Outliers should be identified using rigorous statistical analysis; two methods for this are shown in Chapter 3.

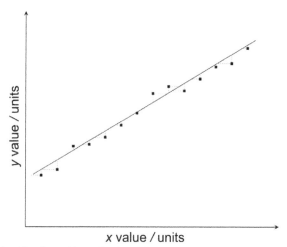

Figure 4.6   The line of best fit should balance the distance of points above and below the line.

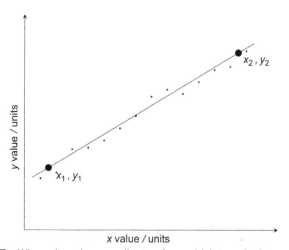

Figure 4.7   When choosing coordinates from which to calculate a gradient, they should lie on the line of best fit and be widely set apart within the bounds of the experimental data.

It is rare to have to draw graphs by hand anymore—except in exams!

The gradient may be determined from knowing two coordinates on a line of best fit. These coordinates should ideally be widely separated within the bounds of the experimental data (Figure 4.7). The further apart these coordinates are, the smaller the uncertainty (otherwise known as 'error') in any calculated value of the gradient or intercept.

The gradient is calculated by determining the 'change in y' and dividing it by the 'change in x'. We use the symbol Δ (the upper-case Greek letter delta) to mean 'change in' a given value:

$$m = \frac{\Delta y}{\Delta x} = \frac{y_2 - y_1}{x_2 - x_1} \qquad (4.9)$$

It is important to remember that the values you read from the graph for x and y belong with their units and any other multiplying term which may be on the axis. Therefore, the gradient of any line has the units of 'units on y/units on x.'

Finally, if the intercept doesn't naturally fall within the range of the drawn graph (remember that the x values may not go all the way to zero!), then it can be calculated by taking one of the values on this line of best fit (one of your coordinate values used for your gradient is fine). Use these (x, y) values by putting it into the equation of the line $y = mx + c$, and use your calculated gradient for m. You can then simply solve this equation to determine the intercept, c. Or, in equation form, from choosing one of the (x, y) points on your line:

$$y - mx = c \qquad (4.10)$$

> You will already be familiar with this as a notation of 'change' in chemistry in equations such as the Gibbs equation, $\Delta G = \Delta H - T\Delta S$.

> Don't forget the units for the gradient.

4.4 Uncertainties in Straight-line Graphs

As with any measured value, calculated values also have uncertainties associated with them. We covered some aspects of error propagation (also called uncertainties) in Chapter 3, but now we turn to the propagation of errors in the context of our graphs. Propagation of errors is simply a fancy term for saying, 'I know the uncertainty in this value but how do I determine the uncertainty in something related to it?' One of the reasons we draw graphs is because measuring a larger number of data points minimises the overall uncertainty in the derived value. For example, in Figure 4.3, the value of the zero-order rate constant, k, can be derived from the gradient of the line of best fit; if we were to simply use a single measurement at time 30 s, measure the concentration at that time, and use that value together with the starting concentration $([A]_0)$ to determine the rate constant k, we would not have any indication of the uncertainty. The benefit of plotting the graph is that we have essentially taken an 'average' of many values (this is not a traditional 'average', as all measurements are taken at different points); this average reduces the uncertainty in exactly the same way as taking any multiple sets of measurements. In fact, you would expect that the more data points make up the data set, the smaller the uncertainty in the line of best fit.

Most data are now analysed using computers, using software tools such as Microsoft Excel and a function called a 'linear regression', or using more specialist software, such as Origin, R or SciPy in Python. As such, it is rare for any such calculations to be done by hand. In this section, we will discuss the principles of what is going on during error analysis so that you are better able to understand what the computer is doing and to recognise whether it is doing it correctly. Of course, it will also help you to do the error analysis manually should the need arise.

4.4.1   *Uncertainties in the Trendline: The Gradient*

To determine the error in the gradient, start with the line of best fit through your data, as in Figure 4.6. Now draw two new lines, each parallel to the line of best fit and at equal distances above and below it. These lines should be set far enough away that two-thirds[‡] of all of your data points are now enclosed within them (see Figure 4.8).

This gives us a pair of lines within which lie two-thirds of our data points; the purpose of this is to help us visualise these new alternative gradient options. These gradient options must be enclosed within our x measurement range, so we define a 'box' following the two parallel gradients between the lowest value of x and the highest value of x, as shown in Figure 4.9. Within this box, the steepest

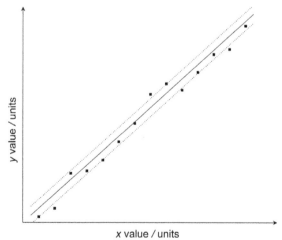

Figure 4.8   When determining the uncertainty in a gradient, the first step is to draw two lines, each parallel to the line of best fit, and sufficiently far apart that $2/3$ of the data points occur between them.

[‡]Technically it is not two-thirds, but instead 68.2%—so that one standard deviation of all data is contained within the box. See Chapter 2 for more insight into the statistical treatment of results.

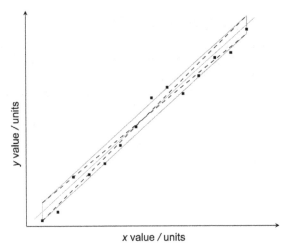

Figure 4.9 Two alternative feasible values of the gradient are presented (dashed lines) within the defined 'box' (parallelogram). These possible alternative gradients run across opposite corners of this box and represent the two extremes.

possible gradient goes from the bottom left corner to the top right, while the shallowest possible gradient goes from the top left to the bottom right.

The 'error' in the gradient is now just the modulus of the difference between the gradient of *one* of these alternative lines and the original line of best fit (as it is a modulus, we can disregard any 'minus' sign):

$$\sigma_m = |m - m_{\min}|$$

or:

$$\sigma_m = |m - m_{\max}|$$

> The modulus (or absolute value) ignores the sign of the value and only concerns itself with the magnitude.

Finally, the corresponding error in the intercept may be determined using a similar method; if the intercept is 'on scale' then the error in the intercept may simply be found by 'reading off' the values of the intercept of the line of best fit and that of one of the alternate gradients, and calculating the difference between these two values. Alternatively, the intercepts for the 'alternative gradients' may be determined in the same method as eqn (4.10). Simply pick a point on the alternate line of best fit:

$$y_{\text{alternative}} - m_{\text{alternative}}\, x_{\text{alternative}} = c_{\text{alternative}} \qquad (4.11)$$

and:

$$\sigma_c = |c - c_{\text{alternative}}|$$

4.5 Plotting Experimental Data

The Beer–Lambert law relates the absorbance of light passing through a sample to the concentration of a species absorbing that wavelength of light. Absorbance is simply a measure of the proportion of light successfully passing through the sample. We use our instruments to measure the difference between the intensity of light passing through a blank sample (containing just solvent) and that passing through the sample.

The Beer–Lambert law itself states:

$$\log\frac{I_0}{I} \equiv -\log\frac{I}{I_0} = A = \varepsilon cl \tag{4.12}$$

where I_0 is the intensity of light passing through the blank; I, the intensity of light passing through the sample; A, the absorbance; c, the concentration of the analyte; l, the distance through the sample the light travels (usually referred to as the 'path length'; it should make sense that the absorbance will be greater if the light has to pass further through the solution) and ε is the 'molar extinction coefficient', a measure of how effectively a molecule absorbs light. Note that A has no units as it is derived from the logarithm of the ratio of intensities (see Chapter 1).

Measuring concentrations using light absorption is extremely useful as the technique is non-destructive (the sample can be recovered) and can be used for small samples, at low concentrations and with continuous flow in a reactor. Therefore, knowing how to analyse the data appropriately is extremely important. To determine the concentration of the analyte, first you need to determine the value for the molar extinction coefficient, ε.

The experimental results in this example are from a titration of a molecule called methylene blue. Small volumes, 2 µL, of a concentrated stock solution of methylene blue (MB) were added and mixed into a 1 cm path length cuvette, initially containing 1200 µL of pure water. The absorbance at 664 nm was recorded after each addition. The methylene blue stock solution contained 50.4 mg of methylene blue ($M_R = 319.85$ g mol^{-1}) dissolved into exactly 1 L (or 1 dm^3) of water.

Starting with the measured experimental data in Table 4.3, the table can be extended to calculate the concentration of the methylene blue in solution, which is shown in Table 4.4.

Table 4.3 Experimental data collected when titrating concentrated methylene blue solution in 2 µL aliquots into 1200 µL of pure water.

V_{added} / µL	$A_{664\ nm}$
0	0.004
2	0.024
4	0.032
6	0.059
8	0.082
10	0.090
12	0.115
14	0.127
16	0.159

Table 4.4 Table extended to show the calculation of the concentration of methylene blue, [MB], in each sample.

V_{added} / µL	V_{total} / µL	[MB] / µmol dm^{-3}	$A_{664\ nm}$
0	1200	0	0.004
2	1202	0.262	0.024
4	1204	0.524	0.032
6	1206	0.784	0.059
8	1208	1.04	0.082
10	1210	1.30	0.090
12	1212	1.56	0.115
14	1214	1.82	0.127
16	1216	2.07	0.159

The concentration of the methylene blue stock solution was calculated from knowing the molar mass, the mass added to the solution and the final volume:

$$n = cv \tag{4.13}$$

$$n = \frac{m}{M_r} \tag{4.14}$$

Combining and rearranging eqn (4.13) and (4.14) gives:

$$c = \frac{m}{M_r v} \tag{4.15}$$

> Try to rearrange with algebra as much as you can before you put any numbers into an equation.

Remember square brackets are used as shorthand for the concentration of a reagent; just putting *concentration* on the axis isn't very informative.

Therefore, the concentration of methylene blue in the stock solution is:

$$\left[\text{methylene blue}\right] = \frac{50.4 \times 10^{-3} \text{ g}}{319.85 \text{ g mol}^{-1} \times 1 \text{ dm}^3}$$

$$\left[\text{methylene blue}\right] = 0.158 \text{ mmol dm}^{-3} \qquad (4.16)$$

The concentration of the methylene blue in each dilution step described in Table 4.3 is calculated by knowing the concentration of the methylene blue stock solution and the initial and final volumes involved in the dilution. From eqn (4.13), the number of moles of methylene blue does not change in a dilution, so:

$$c_{\text{stock}} V_{\text{stock}} = c_{\text{dilute}} V_{\text{dilute}} \qquad (4.17)$$

which rearranges to give:

$$c_{\text{dilute}} = \frac{c_{\text{conc}} V_{\text{conc}}}{V_{\text{dilute}}}$$

We have a c in the general representation and in the Beer–Lambert law; there are only so many symbols, so some inevitably get reused. Make sure that you keep track of the symbols in any given example.

Now we need to plot the graph in order to determine the value of the molar extinction coefficient. Recalling eqn (4.12), we need to think about what we have actively had control over (the independent variable). In this case, we have actively been changing the volume of methylene blue stock solution added (and therefore the concentration) and measuring the resultant absorbance of the solution. So:

$$A = \varepsilon c l \qquad (4.18)$$

and, rearranging slightly to group constants:

$$\begin{array}{ccccc}
y & = & m & x & + & c \\
\downarrow & & \downarrow & \downarrow & & \\
A & = & \varepsilon l & c & &
\end{array} \qquad (4.19)$$

Plotting the concentration of methylene blue on the *x*-axis and the absorbance on the *y*-axis gives a gradient of εl and, in theory, an intercept of 0. However, remember *never to force an intercept through the origin!* (Figure 4.10)

From our graph, we can calculate the gradient, εl, as:

$$\varepsilon l = 0.0724 \text{ }\mu\text{mol}^{-1} \text{ dm}^3$$

Remember that the units are given from the units of *y* divided by the units of *x*.

Since the path length is 1 cm, this gives a value for the molar extinction coefficient as:

$$\varepsilon = 72.4 \times 10^3 \text{ mol}^{-1} \text{ dm}^3 \text{ cm}^{-1}$$

The accepted value of the molar extinction coefficient of methylene blue at 664 nm is 74 000 mol^{-1} dm^3 cm^{-1}.[§] It is normal for

§S. Prahl, 1998, omlc.org.

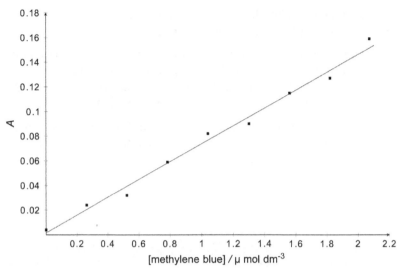

Figure 4.10   Determining the molar extinction coefficient using the Beer–Lambert law.

spectroscopists to leave the units as this (or else report it as $M^{-1}\,cm^{-1}$) because most cuvettes have a path length of 1 cm. It perhaps makes sense to want to combine the units of dm^3 and cm^{-1}, but this is almost never actually done in real life and introduces a possibility for miscommunication through unit conversions.

As mentioned earlier, you should never force a line of best fit through zero. In this case, the value of the intercept (which according to theory should be zero) is 1.5×10^{-3}, which gives a quick 'rough and ready' idea of the errors involved during this experiment.

> Remember, M is sometimes used as shorthand for mol dm^{-3}

4.6 Summary

In this chapter, we have covered the basics of laying out and interpreting tables and graphs in the context of chemistry experiments. A key concept examined is how to lay out units properly for both graphs and tables to avoid any possible ambiguity, namely that units and any common scaling factors (for example $\times 10^2$) are placed behind a solidus (a 'stroke', /) in the table header or axis label.

We also examined the principles behind graphing data, which data set to plot on which axis and how to manipulate a non-linear equation so that we plot appropriate parameters in order to give a straight line on our graphs (which is easier to fit!). We showed this in the context of the Arrhenius equation, but the same principle can be adapted to any situation. Generally, it is unusual for directly measured data to be plotted on a graph, with most data requiring some form of manipulation before plotting.

Plotting graphs offers a way to minimise the effect of erroneous measurements through the inclusion of a line of best fit. The fit-line is also subject to considerations of error analysis, and we have seen a method of determining the uncertainty in a fit-line.

The ability to plot an effective graph is a vital skill for the chemist, and this chapter should give you most of what you need for your pursuit of undergraduate study.

Summary: Chapter 4

Tables

- Units and scaling factors should be placed in the table header behind a solidus ('stroke', /), *not* using brackets; *e.g.* 'Concentration / × 10^4 mol dm'$^{-3}$, *not* 'Concentration (× 10^4 mol dm^{-3})'.
- Data in the main body of the table should be shown as unitless values; *e.g.* '9.43', *not* '9.43 × 10^4 mol dm^{-3}'.
- To convert a value back to its original data quantity from a published table, treat the 'header' and the 'value' as two sides of an equation and rearrange to reform the data; *e.g.* 9.43 = concentration / × 10^4 mol dm^{-3} becomes 9.43 × 10^4 mol dm^{-3} = concentration.

Graphs

- Units and scaling factors should be placed in axis labels behind a solidus in the same manner as for tables.
- To convert a plotted data point back to its original data, again treat the plotted value and the axis label as two sides of an equation, as for tables.
- Plotting a line of best fit offers a route to mitigate any inaccurate measurements made.
- Outlying data points should be identified through statistical treatment, not through appearances.
- A line of best fit can have an uncertainty associated with its gradient.
- Never force a line of best fit to go through the origin, even if a data value was recorded at this point. The difference between the intercept and the origin (if theory predicts a line going through the origin) gives a quick indication of the uncertainty in the experiment.

? Exercise 4.3 Tables and Graphs in a Chemical Context

1. The concentration of a reactant, A, was measured at 60 s interval, as shown in the table. Identify the x values, y values, gradient and intercept in the integrated form of the second-order rate equation (eqn (4.20)), and plot an appropriate graph, determining a value for both $[A]_0$ and k.

$$\frac{1}{[A]} = \frac{1}{[A]_0} + kt \qquad (4.20)$$

t/s	$10^4\,[A]/\text{mol dm}^{-3}$
60	21
120	12
180	8.4
240	7.1
300	5.6

> Don't forget about the factor of 10^4 when you determine the value of k—remember that you can either use it when you calculate values or keep it on the axes.

2. The temperature of a system, T, was varied and the rate of a zero-order reaction monitored. The rate constant at each temperature is given in the table. Rearrange the Arrhenius equation (eqn (4.21)) to a linear form, and plot an appropriate graph to determine both E_a and A.

$$k = Ae^{-E_a/RT} \qquad (4.21)$$

$T/°C$	$k/\text{mol dm}^{-3}\,\text{s}^{-1}$
40	0.00434
50	0.0140
60	0.0421
70	0.118
80	0.316

> Temperatures should be in kelvin!

> Remember that the y-intercept is when the value on x is zero, not just where it crosses the y on your graph!

> k and A will have the same unit.

3. The following data are for a fluorescence-quenching experiment, in which the concentration of a quencher reduces the fluorescence lifetime of a dye. Plot the data so that they fit the Stern–Volmer equation (eqn (4.22)), and determine the rate constant for quenching, k_q, and the 'unquenched' lifetime, τ_0.

$$\frac{1}{\tau} = \frac{1}{\tau_0} + k_q[Q] \qquad (4.22)$$

[Q]/mol dm^{-3}	τ/ns
0.125	1.62
0.250	1.17
0.375	0.99
0.500	0.78
0.750	0.57

4. The following data show how the vapour pressure of diethyl ether changes with temperature. Use eqn (4.23) to plot a graph and determine the enthalpy of vaporisation, $\Delta_{vap} H$, of ether:

$$\ln p = \frac{\Delta_{vap} H}{RT} + C \qquad (4.23)$$

$T/°C$	p/mm Hg
−38	17
−30	28
−25	40
−20	55
−15	75
−10	97
−5	125
0	157

Determine the value of the intercept, C, and hence determine the temperature at which diethyl ether boils under a pressure of 1 atmosphere (760 mm Hg).

5. The decomposition of cyclobutane to ethene proceeds as follows, with the following kinetic data:

$$C_4H_8 \rightarrow 2C_2H_4$$

t/s	$[C_4H_8]$/mol dm^{-3}
0	1.000
10	0.894
20	0.799
30	0.714
40	0.638
50	0.571
60	0.510

t/s	$[C_4H_8]$ / mol dm^{-3}
70	0.456
80	0.408
90	0.364
100	0.326

Plot an appropriate graph to determine the rate constant, k.

Solutions to Exercises

Solutions: Exercise 4.1

1. [HCl] = 0.150 mol dm^{-3}
2. $p = 8.76 \times 10^5$ Pa
3. $E = 316$ mV
4. $K = 4.89 \times 10^4$

5. $\tau = 5.7$ ns
6. $k = 5.2 \cdot 10^{-3}$ dm^3 mol^{-1} s^{-1}
7. $\Delta H = -458.2$ kJ mol^{-1}
8. [A] = 0.040 mol dm^{-3}

> Note that you are asked for a sketch; this means you need to only indicate the shape of the graph and the key characteristics (any intersections, gradients etc.).

Solutions: Exercise 4.2

1.

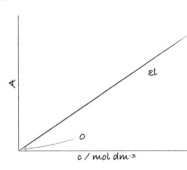

2.

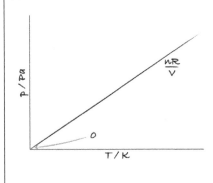

3.

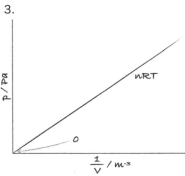

4.

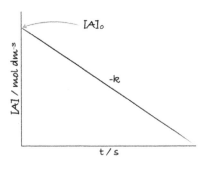

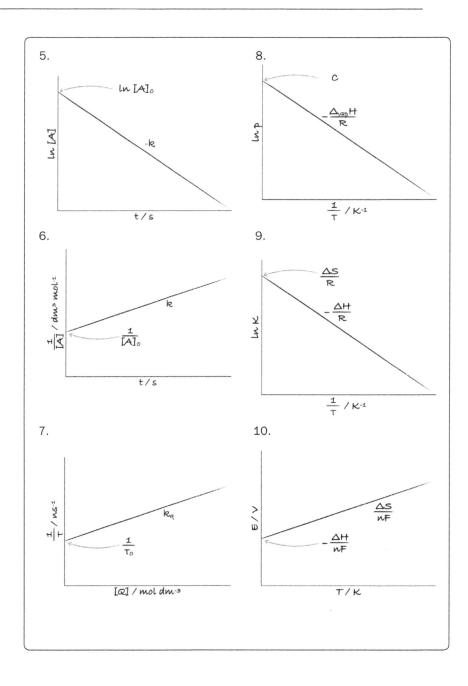

Solutions: Exercise 4.3

In contrast to the sketches in Exercise 4.2, if you are asked to plot data, you should be as accurate as you can, as you will be expected to do calculations using the graph.

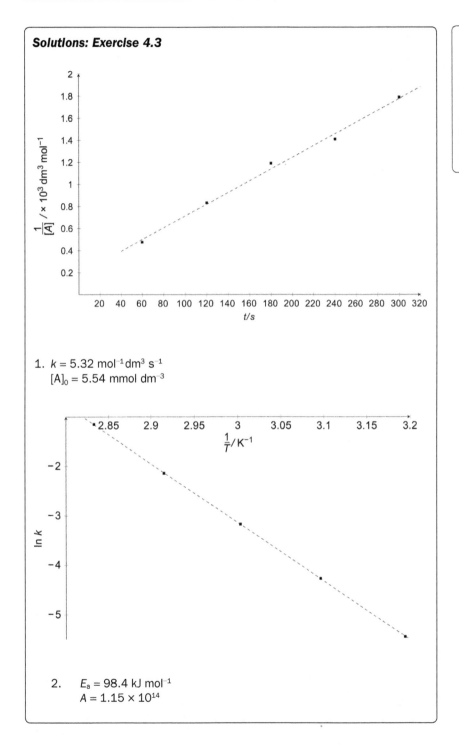

1. $k = 5.32$ mol^{-1}dm^3 s^{-1}
   $[A]_0 = 5.54$ mmol dm^{-3}

2. $E_a = 98.4$ kJ mol^{-1}
   $A = 1.15 \times 10^{14}$

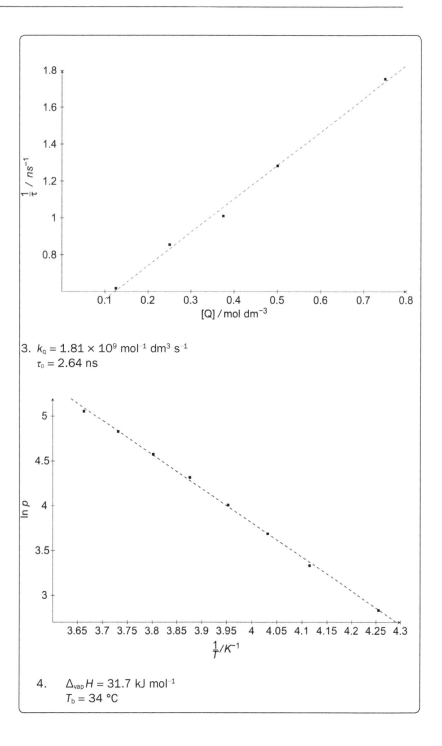

3. $k_q = 1.81 \times 10^9$ mol^{-1} dm^3 s^{-1}
   $\tau_0 = 2.64$ ns

4. $\Delta_{vap}H = 31.7$ kJ mol^{-1}
       $T_b = 34$ °C

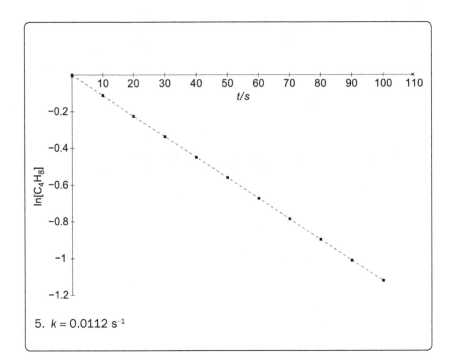

5. $k = 0.0112 \ s^{-1}$

Learning Points: What We'll Cover

☐ Measurement of angles in degrees and radians and interconversion between the two
☐ Elements of trigonometry: the sine, cosine and tangent functions in the context of triangles
☐ Inverse functions of trigonometric functions and the notation used
☐ Application of trigonometric functions to determine unknown quantities, including bond lengths and bond angles
☐ Cartesian and polar coordinate systems and interconversion between the two
☐ Applications of polar coordinates within chemistry

Why This Chapter Is Important

• The shapes of molecules are fundamental to understanding their properties; trigonometry allows us to do calculations on the shapes of molecules.
• Trigonometry allows us to build molecular models which can accurately predict molecular properties and interactions.
• Trigonometric functions are not just used in geometry; they are found in all areas of mathematics, and will crop up in further chapters as we continue to apply maths in a chemistry context.

Trigonometry and Coordinate Systems: Describing Molecular Positions

Triangles are everywhere when we consider three-dimensional models; from geographical mapping to computer gaming, the triangle is ubiquitous in calculating distances and separations. In chemistry, we use them extensively in determining molecular geometries, bond angles and intermolecular separations, so understanding the mathematics of trigonometry is essential to creating functional chemical models.

The principles of trigonometry have wider applications in describing the shapes of waves; this becomes essential when we start to consider wave functions in quantum chemistry and spectroscopy. In these circumstances, sine and cosine functions move away from direct triangular relations, instead demonstrating their cyclic 'wave' nature, making them ideal for mathematical descriptions of waves. In this chapter, we introduce the application of trigonometry when considering how liquids 'wet' solid surfaces, as well as the model used to determine scattering angles which ultimately explains why the sky is blue. The principles of trigonometry introduced in this chapter will continue to be used throughout this text.

5.1 Determination of Molecular Distances

We can determine separations within molecules by identifying triangles and using trigonometry to determine all other otherwise unknown dimensions of the triangle, and hence of the molecule. This is effectively a molecular equivalent of the Ordnance Survey's Principal Triangulation of Great Britain, in which land heights around the country were identified by measuring the angles between defined points. The triangle is effectively the simplest surface which can be modelled in a computer; in the early days of computer graphics, a processor's speed was often touted in 'triangles per second'.

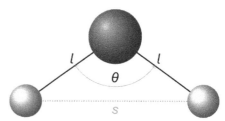

Figure 5.1   The structure of a water molecule. If we know the bond angle θ and each bond length l, we are able to calculate the separation s between the two hydrogen atoms.

In chemistry, we continue to examine triangular properties in respect of molecular dimensions, but also more widely in vectors, in polar coordinates and in computational chemistry. An understanding of triangles and of the resulting *trigonometric functions* is vital for our continued study of molecular geometries.

Consider the water molecule, H_2O (Figure 5.1); if we know the O–H bond length and the bond angles, we can determine the separation between the hydrogen atoms. We can also calculate additional factors based on the molecular parameters in H_2O, such as the vertical height of the molecule, h, and, by knowing the relative masses of the atoms, we can determine the position of the centre of mass of the molecule.

To determine these parameters, we need to know the basic relationships in trigonometry, as well as introduce a new angular concept: the *radian*.

5.1.1 Elements of Trigonometry: Angular Measurement

You are already probably familiar with the idea of recording angles in degrees. There are two axioms (unprovable truths) when considering measurement of angles:

1. There are 360° in a full rotation.
2. There are 180° on a straight line.

From these two rules, we can construct almost all of our geometry when working with degrees. However, the vast majority of trigonometry uses angles measured in *radians*. The radian is a curious unit; there is not a whole number of radians in a full rotation. Because of its definition (Key Concept 5.1), by knowing how many times the radius, r, fits into the circumference, we can find out how many radians are in a full rotation. The circumference of a circle is defined as $2\pi r$; dividing this by the radius gives us 2π radians in a full rotation. This gives us some insight as to why we often express radians as a fraction of π; this is then a fraction of the whole rotation.

By knowing that there are 360° in a full rotation, and $2\pi^c$ in a full rotation, we can determine a conversion factor, as:

$$\text{Degrees per radian}: \quad \frac{360°}{2\pi\,\text{rad}} \quad \approx \quad 57.3°\,\text{rad}^{-1} \qquad (5.1)$$

> ^{c} or 'rad' are both used as units of radians.

> ### ❶ Key Concept 5.1 The Radian as a Measurement of Angle
>
> The radian is defined as *'the angle which forms an arc of equal length to the radius of the circle'*.
>
> Graphically, this is illustrated as:
>
>
>
> The radian is an extremely important unit in mathematics, finding wide use across a large number of applications. In chemistry, it finds application in spectroscopic considerations, as well as quantum mechanics.
>
> The symbol for 'radians' is a superscript c, and an angle in radians is often expressed as a fraction of π; for example $2\pi^c/_3$.

❓ Exercise 5.1 Conversion of Angles

Use the conversion factor in eqn (5.1) to convert the following angles as described.

Convert the following values in degrees to radians:

1. 532°
2. 187°
3. 77°
4. 624°
5. 175°
6. 241°

Convert the following values in radians to degrees:

7. 0.18^c
8. 3.57^c
9. 1.03^c
10. 7.79^c
11. 3.60^c
12. 3.81^c

5.1.2 Elements of Trigonometry: The Triangle

Any flat triangle can be completely described by any one of the following sets of conditions:

1. All three side lengths
2. Any two sides and one angle
3. One side and any two angles

Any one of these sets will specify *congruent* triangles (triangles completely identical in size and shape), while specification of just three angles will give *similar* triangles (the same shape, but not necessarily with the same size).

5.1.3 Elements of Trigonometry: The Right-angled Triangle

The *right-angled triangle* shows the key trigonometric functions which allow us to do geometric calculations, as well as a key theorem in triangle geometry: Pythagoras' theorem (see Toolkit A.10). An example of a right-angled triangle is shown in Figure 5.2, where the labelling of the sides relative to an angle of interest is demonstrated.

5.1.4 Elements of Trigonometry: SOH-CAH-TOA

The right-angled triangle in Figure 5.2 allows us to define the key trigonometric functions of the angle θ, namely *sine, cosine* and *tangent*. These are summarised in eqn (5.2); note that the tangent function is not only a ratio of the sides, but also a ratio of the sine and cosine functions of the angle.

$$\text{sine:} \quad \sin\theta = \frac{\text{opposite}}{\text{hypotenuse}}$$

$$\text{cosine:} \quad \cos\theta = \frac{\text{adjacent}}{\text{hypotenuse}}$$

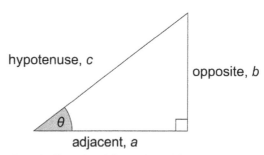

Figure 5.2 The labelling of a right-angled triangle. The side labels 'opposite' and 'adjacent' are always relative to the angle θ.

$$\text{tangent:} \quad \tan\theta = \frac{\text{opposite}}{\text{adjacent}} \equiv \frac{\sin\theta}{\cos\theta} \qquad (5.2)$$

Each of these functions specifies the ratio of side lengths required to give a particular angle; for example, a triangle containing an angle of 60°, ($\pi/3^c$), the ratio of the adjacent side to the hypotenuse must be 0.5 (*i.e.* cos(60°) = 0.5).

Using sine, cosine or tangent functions allows us to determine an unknown side length of any triangle, either through direct application to a right-angled triangle, as shown in Figure 5.2, or through the subdivision of any other triangle into right-angled triangles.

These trigonometric functions have considerably deeper meaning beyond merely triangulation, and we will continue to use them in future chapters.

> Remember that the symbol for a radian is a superscript 'c' and that we usually express radians as a fraction of π where possible, as it is impossible to express π exactly in any other way.

❓Exercise 5.2 Calculation of Side Lengths from Angle Measurements

---

Considering the triangle layout shown in Figure 5.2, calculate the required side length from the given values.

1. Find a; $c = 0.37$ cm; $\theta = 73°$.
2. Find b; $c = 1.41$ cm; $\theta = 82°$.
3. Find c; $b = 2.11$ cm; $\theta = 81°$.
4. Find c; $b = 0.11$ cm; $\theta = 13°$.
5. Find a; $c = 3.45$ cm; $\theta = 12°$.
6. Find c; $a = 0.81$ cm; $\theta = 59°$.
7. Find b; $a = 1.68$ cm; $\theta = 74°$.
8. Find a; $b = 1.73$ cm; $\theta = 71°$.

5.1.5 In Context: Contact Angles at Interfaces

Whenever there is a liquid on a solid surface where both surfaces are exposed to the air (a gas), the liquid 'wets' the surface and there is a defined *contact angle*, θ_c, at the liquid–solid interface for a given temperature and pressure. This is illustrated in Figure 5.3, which shows how the contact angle is defined between the 'wet' surface

(a) A droplet considered to 'wet' the surface.

(b) A droplet that does not 'wet' the surface.

Figure 5.3   The contact angle θ of a liquid wetting a surface. The angle is defined as the angle between the surface and the tangent to the liquid surface at the point where the three phases (solid, liquid and gas) meet. (a) Strong interactions between the solid and liquid phases resulting in wetting of the surface (a hydrophilic surface); (b) weaker interactions and poor wetting (a hydrophobic surface).

and the tangent to the liquid's surface at the point where the solid, liquid and gas phases meet.

The contact angle depends on the surface tensions (also called interfacial energies) between the solid and liquid, γ_{SL}, the solid and gas γ_{SG} and liquid and gas γ_{LG}, as:

$$\gamma_{SG} = \gamma_{SL} + \gamma_{LG}\cos\theta \qquad (5.3)$$

Eqn (5.3) also accounts for the angle of the observed meniscus as liquid rises in a capillary tube (Figure 5.4). In capillary rise experiments, the surface tension of the liquid–gas interface is given by eqn (5.4), where h is the height of the capillary rise from the surface of the liquid to the bottom of the meniscus in the capillary tube, r the internal radius of the capillary tube, g the acceleration due to gravity ($g = 9.80665$ m s^{-2} as a standard value) and ρ_L and ρ_G the densities of the liquid and gas, respectively:

$$\gamma_{LG} = \frac{1}{2}gh(\rho_L - \rho_G)\frac{r}{\cos\theta_c} \qquad (5.4)$$

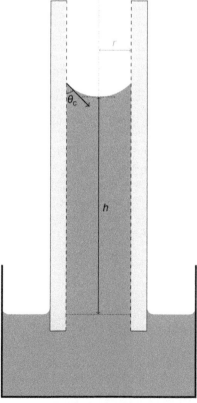

Figure 5.4   The contact angle θ_c of a liquid wetting a surface during capillary rise obeys the same laws of thermodynamics as that of a droplet on a surface.

The contact angle in eqn (5.4) is often assumed to be (or approximated to be) $0°$, such that $\cos\theta = 1$.

5.1.6 Elements of Trigonometry: Inverse Functions

We have seen how to calculate an unknown side from knowing the angles in a triangle; however, a much more common application (particularly in chemistry) is to use the ratio of the sides to determine what the unknown angle might be.

As we discussed in Chapter 1, every function must have an inverse, and trigonometric functions are no exception. There are two forms of notation for these: $\sin^{-1}(x)$ and $\arcsin(x)$. These are shown in Table 5.1, along with the notation for the inverse cosine and inverse tangent functions.

At this point, it is worth highlighting the conflict of notation between 'functions' and 'powers'. While $\sin^{-1}(x)$ is the inverse function and uses the same notation as '$f^{-1}(x)$' (that is, the function which 'undoes' the sine function; see Chapter 1), $\sin^2(x)$ is a 'power function', meaning the same as $(\sin(x))^2$. It is important to recognise that $\sin^{-1}(x)$ *is not the same* as $1/\sin(x)$. Instead, the alternate terms *cosecant* (cosec), *secant* (sec) and *cotangent* (cot), shown in Table 5.2, are used to show these reciprocal functions.

> There is competing (and contradictory) notation with trigonometric functions. Remember that $\sin^{-1}(x)$ *is not the same* as $1/\sin(x)$.

Table 5.1 Notation for inverse trigonometric functions. It is worth recognising that the example $\sin^{-1}(x)$ is the same way of notating an inverse function as '$f^{-1}(x)$' and should be thought of in the same manner.

$\sin\alpha = x$	$\cos\beta = y$	$\tan\gamma = z$
$\alpha = \sin^{-1}(x)$	$\beta = \cos^{-1}(y)$	$\gamma = \tan^{-1}(z)$
$\alpha = \arcsin(x)$	$\beta = \arccos(y)$	$\gamma = \arctan(z)$

Table 5.2 Notation for reciprocal trigonometric functions. These reciprocals cannot be represented as $\sin^{-1}(x)$, as this notates an inverse function, as shown in Table 5.1.

Reciprocal function		Notation
$\dfrac{1}{\sin(\alpha)}$	$\equiv$	$\operatorname{cosec}(\alpha)$
$\dfrac{1}{\cos(\beta)}$	$\equiv$	$\sec(\beta)$
$\dfrac{1}{\tan(\gamma)}$	$\equiv$	$\cot(\gamma)$

? Exercise 5.3 Calculation of Angles

With reference to the right-angled triangle in Figure 5.2, calculate the angle θ from the two given sides.

1. $a = 1.91$ cm; $c = 2.14$ cm.   5. $a = 3.26$ cm; $b = 1.66$ cm.

2. $b = 0.28$ cm; $a = 0.91$ cm.   6. $b = 0.11$ cm; $a = 0.13$ cm.

3. $c = 4.21$ cm; $a = 3.41$ cm.   7. $b = 1.47$ cm; $c = 3.47$ cm.

4. $c = 3.91$ cm; $b = 2.19$ cm.   8. $c = 2.15$ cm; $b = 2.13$ cm.

5.1.7 In Context: Rayleigh Scattering of Light

As light travels, it is scattered by any particles in its path. Small particles (such as the gas molecules in our atmosphere) scatter light such that the sky appears blue in the day and red at sunset, while the larger particles of soot and other pollutants enhance this scattering; this is why particularly beautiful sunsets are often seen during periods of high atmospheric pollution.

The angle, θ, through which light is scattered depends on a number of factors; however, the principal factor is the wavelength of light, λ. Longer wavelengths (red light) are scattered less than shorter wavelengths (blue light). The intensity of the light scattered through an angle θ is given by eqn (5.5); N is the number of scattering particles in the path of the light, α the polarisability of the particle, r the distance of the observer from the scatterer, and I and I_0 the measured and incident intensity, respectively:

Recall that $\cos^2\theta$ means $(\cos\theta)^2$.

$$I = I_0 \frac{8\pi^4 N\alpha^2}{\lambda^4 r^2}\left(1+\cos^2\theta\right) \tag{5.5}$$

When the sun is low on the horizon, it passes through a larger distance to reach the observer; as a result of this long path through our atmosphere, the blue light has been fully scattered away from the observer, leaving only the red light from the sun being observed.

A number of other models describe light-scattering processes, and are used variously for scattering by different particle sizes; however, the principles governing the angle of scatter (θ) of incident light remain the same in all models.

5.1.8 Elements of Trigonometry: Standard Results

Within trigonometry, it is useful to be aware of standard results, and these can be visualised with the aid of two triangles, shown in Figure 5.5.

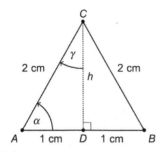

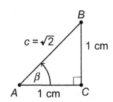

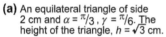

(a) An equilateral triangle of side
2 cm and $\alpha = {}^{\pi}/3$, $\gamma = {}^{\pi}/6$. The
height of the triangle, $h = \sqrt{3}$ cm.

(b) An isosceles right-angled triangle
of side 1 cm and $\beta = {}^{\pi}/4$. The
hypotenuse of the triangle, $c = \sqrt{2}$ cm.

Figure 5.5   Using two triangles, one equilateral, the other isosceles, we
can determine values of trigonometric functions. The height of
the first triangle and the hypotenuse of the second are found
using Pythagoras' theorem.

Table 5.3   Standard trigonometric results for ${}^{\pi}/3{}^{c}$, ${}^{\pi}/4{}^{c}$ and ${}^{\pi}/6{}^{c}$.

Function	${}^{\pi}/2{}^{c}$ (90°)	${}^{\pi}/3{}^{c}$ (60°)	${}^{\pi}/4{}^{c}$ (45°)	${}^{\pi}/6{}^{c}$ (30°)	0
$\sin(\theta)$	1	$\sqrt{3}/2$	$1/\sqrt{2}$	$1/2$	0
$\cos(\theta)$	0	$1/2$	$1/\sqrt{2}$	$\sqrt{3}/2$	1
$\tan(\theta)$	∞	$\sqrt{3}/1 = \sqrt{3}$	$1/1 = 1$	$1/\sqrt{3}$	0

The triangles shown in Figure 5.5 allow us to determine standard results for the trigonometric functions acting on angles of
${}^{\pi}/3{}^{c}$, ${}^{\pi}/4{}^{c}$ and ${}^{\pi}/6{}^{c}$ shown in Table 5.3.

It is useful to be able to recall the results in Table 5.3, though not
essential. It does, however, illustrate the complementary nature of
the sine and cosine functions of the angles.

5.1.9   Sine and Cosines of Angles Greater Than 90°

The sine and cosine functions have a use and meaning outside the
consideration of triangles. The functions themselves are *cyclic,*
meaning that they repeat with a defined period. This is shown in
Figure 5.6.

The fact that these functions are cyclic means that we can determine a sine or a cosine of any angle, regardless of whether or not
it fits in a triangle. This will become very useful when considering
angles within coordinate systems (Section 5.3.2), when considering

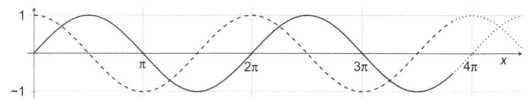

Figure 5.6 Plotting a graph of sin(x) (solid) and cos(x) (dashed) shows the repeating cyclic nature of the functions. For the functions as written, the period of the cycle is 2π; *i.e.* the cycle repeats after every interval of 2π; or, mathematically, sin(x) = sin(x + 2π) and cos(x) = cos(x + 2π).

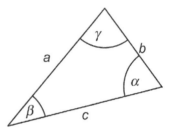

Figure 5.7 A general triangle to demonstrate the principles of the cosine rule and the sine rule. Note that, by convention, sides are labelled in Latin letters (*a*, *b* and *c*), while angles are labelled in Greek letters (α, β and γ), with angle α opposite *a*, β opposite *b* and so on.

vector algebra (Chapter 6), and even when considering complex numbers (Chapter 9).

Returning briefly to triangles; the sines and cosines of angles greater than 90° also find use in the *sine and cosine rules*. These are mathematical formulae which give insight into the connection between a triangle and its sides and angles.

The Cosine Rule

This is most often used in chemistry for finding interatomic separations in molecules. If we consider a general triangle (Figure 5.7), with sides *a*, *b* and *c* and angles α, β and γ, we can look at the relationships present in the cosine rule.

Provided we know two sides and the angle between them, we can calculate the length of the third side of the triangle using the cosine. The general form of the cosine rule is:

$$c^2 = a^2 + b^2 - 2ab\,\cos(\gamma) \tag{5.6}$$

The cosine rule is a relative relationship, and as such the expression in eqn (5.6) may be rewritten for the other two sides of the triangle in Figure 5.7, as shown in eqn (5.7):

$$a^2 = b^2 + c^2 - 2bc \cos(\alpha)$$
$$b^2 = a^2 + c^2 - 2ac \cos(\beta) \tag{5.7}$$

If we know all the side lengths, we can calculate any angle by re-arranging the cosine rule.

The Sine Rule

Referring to Figure 5.7, we can express a relationship between the side lengths for the sines of the respective angles. This is shorter than the cosine rule and is shown in eqn (5.8); however, this is of less direct use in chemistry and is included here for completeness:

$$\frac{\sin \alpha}{a} = \frac{\sin \beta}{b} = \frac{\sin \gamma}{c} \tag{5.8}$$

Once again, a rearrangement allows us to find side lengths or angles depending on what combination of angles and sides we already know.

Example 5.1 Using the Cosine Rule to Determine Interatomic Separations

Q: Tetrachloromethane (CCl_4) is a tetrahedral molecule with a bond angle of 109.5° and C–Cl bond length of 1.75 Å. Calculate the distance between chlorine atoms in this molecule.

A: We can start to solve this problem with a quick sketch.

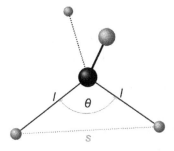

Within this sketch, we have visualised a triangle with two chlorine atoms and the carbon atom at its points. We know that the Cl–C–Cl bond angle, θ, is 109.5° and that the two sides forming this are of length $l = 1.75$ Å.

Applying the cosine rule, we can determine that the separation, s can be found from:

$$s^2 = l^2 + l^2 - 2ll \cos(\theta)$$
$$= 2l^2 - 2l^2 \cos(\theta)$$
$$= 2l^2 \left(1 - \cos(\theta)\right)$$
$$= 2 \times \left(1.75 \text{ Å}\right)^2 \times \left(1 - \cos(109.5°)\right)$$
$$= 8.17 \text{ Å}^2$$
$$s = \sqrt{8.17 \text{ Å}^2} = 2.86 \text{ Å}$$

?Exercise 5.4 Further Calculations of Triangles

Each of the following examples will give you three pieces of information about a triangle; referring to the triangle representation in Figure 5.7, calculate the specified unknown side or angle.

Application of cosine rule:

1. $a = 1.75$ cm, $b = 2.38$ cm, $\gamma = 35°$; calculate c.
2. $b = 3.56$ cm, $c = 6.48$ cm, $\alpha = 18°$; calculate a.
3. $a = 1.74$ cm, $c = 2.65$ cm, $\beta = 113°$; calculate b.

The next two examples will require a more advanced rearrangement of the cosine rule:

4. $a = 0.67$ cm, $b = 1.72$ cm, $\alpha = 22°$; calculate c.
5. $a = 0.84$ cm, $c = 3.33$ cm, $\gamma = 63°$; calculate b.

Application of sine rule:

6. $a = 2.64$ cm, $b = 2.81$ cm, $\alpha = 22°$; calculate β.
7. $b = 1.83$ cm, $\beta = 67°$, $\gamma = 78°$; calculate c.
8. $a = 0.43$ cm, $c = 2.45$ cm, $\alpha = 9°$; calculate γ.

Application of either the sine or the cosine rule:

9. $a = 1.86$ cm, $b = 2.70$ cm, $c = 2.36$ cm; calculate γ.
10. $b = 3.34$ cm, $c = 2.60$ cm, $\gamma = 50.0°$; calculate β.
11. $a = 2.58$ cm, $\alpha = 30.0°$, $\gamma = 100°$; calculate c.
12. In ethanal (CH_3CHO), the aldehyde group ($-CHO$) has an OCH bond angle of $118.6°$ and bond lengths of 1.216 Å (C–H) and 1.114 Å (C=O). Calculate the separation between the carbon and hydrogen centres.[†]

5.2   Molecular Geometry: Molecular Spectroscopy

Being able to determine angles in molecules is extremely important in the consideration of molecular transformation and energy transitions. One example of this is in *rotational spectroscopy* (microwave spectroscopy), concerning the kinetic energy of rotation of the molecule.

The general principles of molecular rotation are:

1. The molecule can rotate around the perpendicular x-, y- and z-axes independently.

[†]Data from H. Hollenstien and H. H. Gunthard, *Spec. Acta.*, 1971, **27A**, 2027.

2. Axes of rotation *always* go through the centre of mass of the molecule.
3. *By convention,* the z-axis is usually considered to be 'vertical' and the primary rotation axis to hich others are referred (this will be discussed when you study molecular symmetry).
4. Any rotation can be simplified to a combination of rotations about each of the *x-*, *y-* and *z-*axes.

To quantify the rotational kinetic energy of a molecule, we need to identify the *perpendicular distance* of each atom from the defined *axis of rotation* (the shortest distance from the axis). In some situations, we can directly use the bond length, as this is already the perpendicular distance. For example the octagonal SF_6 molecule, Figure 5.8, defining the axes will be largely arbitrary owing to the symmetry of the molecule.

The reason for knowing this perpendicular distance is that the 'moment of inertia' of a rotating molecule is key in determining its energy of rotation. This may be thought of as a 'torque'; the further a mass rotates from an axis, the more 'torque' ('twisting power') the mass has. To put it another way, the farther the mass is from the axis of rotation, the harder it will be to stop it rotating. (Think about how difficult it is to stop a large rotating bicycle wheel with your hands while changing a tyre, compared with a small bicycle wheel. Generally, there will only be a small difference in the mass of the wheels, but the larger wheel will be much harder to stop spinning!) An example calculation of this perpendicular distance for water is given in Example 5.2.

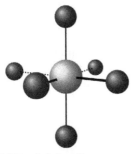

Figure 5.8   Structure of SF_6; defining the axes as *x, y* or *z* is largely arbitrary, owing to the highly symmetrical nature of the molecule; perpendicular distances from the principal axis are simply equal to the bond length.

Example 5.2 Molecular Geometry of H_2O

When we consider a water molecule, we imagine it in three-dimensional space.

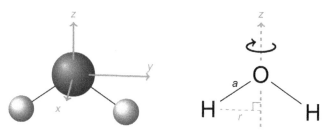

By convention, we orientate the molecule so that the z-axis is the principal axis of the molecule. For now, we will consider rotation about this principal axis.

We know that the bond length a = 0.96 Å and that the HOH bond angle, α, is 104.50°. To find the perpendicular distance of the hydrogen atoms from the axis of rotation, r, we simply apply the trigonometric functions introduced in Section 5.1.4. However, as the principal axis bisects the bond angle, we will use the angle 104.50° ÷ 2 = 52.25°.

Looking at the trigonometric functions relative to the angle: r is opposite the angle, while a is the hypotenuse of the right-angled triangle. Therefore, we can easily see that we need to use a sine function to determine the perpendicular distance:

$$r = a\sin\left(\frac{\alpha}{2}\right)$$

$$= 0.96\,\text{Å} \times \sin\left(\frac{104.50°}{2}\right)$$

$$= 0.76\,\text{Å}$$

This perpendicular distance can then be used to calculate the moment of inertia, $I = mr^2$ for rotational spectroscopy.

？Exercise 5.5 Molecular Geometries

For each of these molecules, determine the perpendicular distance shown.

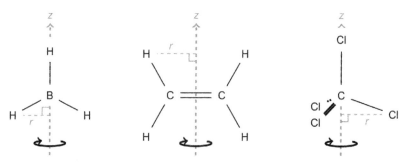

1. For BH_3, the HBH bond angle is 120° and the B–H bond length is 1.19 Å.
2. For C_2H_4, the HCH bond angle is 117°, the C–H bond length is 1.08 Å and the C=C bond length is 1.34 Å.
3. For CH_4, the HCH bond angle is 109.5° and the C–H bond length is 1.09 Å.

For each of the examples, there is more than one way to solve the problem; whichever way you approach the problem, you should still arrive at the same result.

5.2.1 Double-angle Formulae

Another set of trigonometric identities which is useful to know is the set of *double-angle formulae*. These identities allow us to determine the sine and cosine values of pairs of adjacent angles and (more importantly) allow us to use substitutions in our calculations. These will find use in integration (see Chapter 8), as they allow us to simplify a problem into fewer components. One example is that it is much simpler to consider a formula as $\sin(2\theta)$ than as $2\sin(\theta)\cos(\theta)$; this is a valid substitution as these are identical expressions. The double-angle formulae are listed in Key Concepts 5.2 and 5.3.

---

❶ Key Concept 5.2 Double-angle Formulae: Equal Angles

Double-angle formulae offer a way to combine trigonometric functions into a simpler expression. Possible double-angle formulae are:

$$\sin(2\theta) = 2\sin(\theta)\cos(\theta)$$
$$\cos(2\theta) = \cos^2(\theta) - \sin^2(\theta)$$
$$= 2\cos^2(\theta) - 1$$
$$= 1 - 2\sin^2(\theta)$$
$$\tan(2\theta) = \frac{2\tan(\theta)}{1 - \tan^2(\theta)}$$

---

❶ Key Concept 5.3 Addition Angle Formula: Unequal Angles

When angles are not equal to each other, the identities change to the following results:

$$\sin(A+B) = \sin(A)\cos(B) + \sin(B)\cos(A)$$
$$\cos(A+B) = \cos(A)\cos(B) + \sin(A)\sin(B)$$

$$tan(A + B) = \frac{tan(A) + tan(B)}{1 - tan(A)tan(B)}$$

These can be extended if there are more than two angles to consider; for example, for $sin(A + B + C)$:

$$sin(A + B + C) = sin([A + B] + C)$$
$$= sin(A + B)cos(C) + sin(C)cos(A + B)$$
$$= ([sin(A)cos(B) + sin(B)cos(A)] \times cos(C))$$
$$+ (sin(C) \times [cos(A)cos(B) + sin(A)sin(B)])$$

5.3 Coordinate Systems

When calculating molecular properties, we construct a model which accurately represents the geometry of the molecule of interest. To do this, we assign coordinates to each of the atoms in a molecule and plot them on appropriate axes. We need to be aware of two coordinate systems: *Cartesian coordinates*, using x-, y- and z-axes, and *spherical polar coordinates*, using a radius r and two angles, ϕ and θ. You will meet both of these in the course of your chemistry studies so it is worth going over some details about each and how to interconvert between the two.

There is also a system of *cylindrical polar coordinates*, which uses a radius r, a height h and an angle ϕ; this is rarely used in chemistry and only mentioned here for completeness.

The key point with all coordinate systems is that the coordinates must be completely independent of each other; we use the term 'orthogonal' to describe this (see Chapter 6 for more detail on orthogonality), but you can use the term 'perpendicular' when describing simple systems.

5.3.1 Cartesian (x, y, z) Coordinates

You are most likely to be familiar with Cartesian coordinates, using x- and y-axes to plot a great many graphs. The x- and y-axes are independent of each other and, in this geometry, their 'orthogonality' means that the axes are perpendicular to each other.

Cartesian coordinates tell us how far to go along each axis: the coordinate (4, 3) means, 'Go four steps in the direction of the x-axis, and then up three steps in the direction of the y-axis' (Figure 5.9).

This is a convenient way of working when we plot graphs (see Chapter 4); it allows us to visualise a wide range of data, including rates of reaction in our chemical systems (where we plot the variation in concentration with time, as in Figure 5.10). We can also plot in three dimensions to visualise multivariate systems in three dimensions, *e.g.* showing how pressure, volume and temperature interact (Figure 5.11).

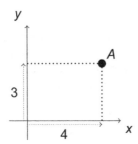

Figure 5.9   Plotting point *A* in Cartesian coordinates. Point *A* is represented by the coordinates (4, 3).

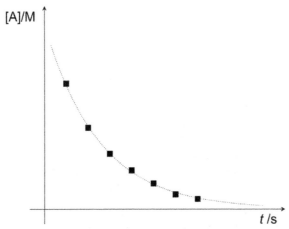

Figure 5.10   Reaction rate data showing the decay in concentration of
reactant A with time. This is an example of a two-dimensional
Cartesian plot, where concentration is plotted as a function
of time.

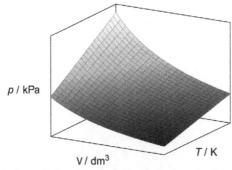

Figure 5.11   A three-dimensional surface showing the interrelation of the
pressure, volume and temperature terms in the ideal gas
equation. For each of the volume and temperature conditions
imposed on a system, the pressure is recorded, giving us our
third coordinate.

These plots suit us very well; however, when we start to visualise chemical phenomena such as in rotating molecules or atomic orbitals, they rapidly become cumbersome. To better represent our models of some chemical phenomena, it makes sense to introduce an alternative coordinate system.

5.3.2   Polar (r, ϕ, θ) Coordinates

Polar coordinates are a way of expressing a position in terms of a distance from an origin (radius, r), and an angle, called the *azimuth*, ϕ. If we wish to discuss positions in three dimensions (*spherical* polar coordinates), we add a second angle, the *polar* angle, θ. In the case of polar coordinates, the three terms cannot be considered perpendicular to each other, however they are still *orthogonal* as they are mutually independent.

You will already be familiar with a form of spherical polar coordinates; whenever you look at a map reference or consider longitude or latitude in the context of global positioning, you will have considered position in terms of two angles. Greenwich, London, for example has the coordinates (51.48 °N, 0.00 °E), while Greenwich Village, New York, has the coordinates (40.73 °N, 74.0 °W). This is a form of spherical polar coordinates; we ignore the radial term as the two points both sit on the same surface (the Earth's surface!), while the angles tell us how far north to go and how far east or west to go.

Consider the Cartesian system in Figure 5.9; the point A has been expressed in terms of distances along the x- and y-axes. We can change this to express point A as a distance from the origin and an angle ϕ (Figure 5.12).

Expressing some systems in polar coordinates can simplify our calculations. When considering the rotation of molecular systems, we could express a position in terms of x and y; however, these are not independent. When x varies, y must also vary in a specific manner

The symbols ϕ and θ, respectively, have a vertical and a horizontal line. The vertical line in ϕ can be thought of as 'symbolising longitude', so ϕ is the *azimuth* angle, while the horizontal line in θ can be thought of as 'symbolising latitude' between the poles, so θ is the *polar* angle.

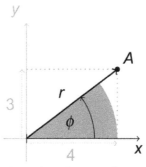

Figure 5.12   Plotting point A in polar coordinates. Point A is now represented by the coordinates (5, 0.64); take 5 steps from the origin and rotate by an angle of 0.64ᶜ (0.64 rad). Note that, by convention, polar coordinates give angles in *radians*. Here, 0.64ᶜ corresponds to 36.9°.

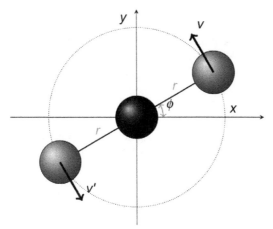

Figure 5.13 The oxygen atoms in a rotating carbon dioxide molecule have their positions defined by a set of x and y coordinates; these are 'locked together', such that $x^2 + y^2 = r^2$, defining their circular motion. Since r is a constant in circular motion, we can express its motion in terms of a single variable; angle ϕ.

in order to keep the molecule on a circular track. This interrelation is shown in Figure 5.13.

We describe such systems as having 'degrees of freedom'; namely 'variables which can change independently and still have the same system'. Although the equation for circular motion, $x^2 + y^2 = r^2$, has two variables, x and y, if x varies, then y can can only take certain values if r is to remain a constant. Since this means that there is only one degree of freedom, we can simplify our mathematics a great deal by reworking the problem to have a single variable, namely the angle ϕ.

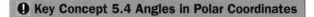

❗ Key Concept 5.4 Angles in Polar Coordinates

Polar coordinates specify a position in terms of a radius and two angles. Imagine a hydrogen atom with its electron at distance r.

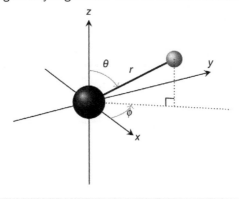

This image relies on a classical interpretation of the atom which is regarded as incorrect, but it provides one way of thinking about spherical polar coordinates.

The electron is at distance r from the nucleus, and it is defined as a direction in terms of azimuth (rotation, ϕ) and polar angle (θ). We do need to note the limits on the angles:

r	Any value
ϕ	$0 \leq \phi \leq 2\pi$
θ	$0 \leq \theta \leq \pi$

Note the difference between the limits: for ϕ, a full rotation is permitted, while for θ only a half rotation is permitted. Think about this in terms of specifying a global position; we can have a full rotation around the equator, but we can only have a half-rotation from north to south. If we permit a full north–south rotation, we end up going past the South Pole and coming up the other side of the planet!

---

❶ Key Concept 5.5 Converting from Polar to Cartesian Coordinates in Two Dimensions

In two dimensions, the conversion from Cartesian to polar coordinates has a direct relation to the trigonometric principles given in Section 5.1.4. The x, y and r components all form a right-angled triangle, with rotation angle ϕ within that triangle.

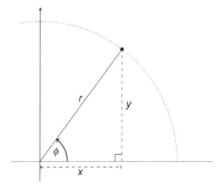

We can then apply the trigonometric identities to determine expressions for x and y in terms of r and ϕ (see Section 5.1.4)

$$x = r\cos(\phi)$$
$$y = r\sin(\phi)$$

> ### ❗ Key Concept 5.6 Converting from Polar to Cartesian Coordinates in Three Dimensions
>
> In three dimensions, things are slightly trickier, but the central trigonometric principles still apply. Remember that ϕ is the angle in the x, y plane, while θ is the angle from the positive z-axis.
>
>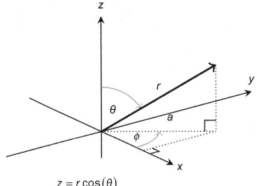
>
> $$z = r\cos(\theta)$$
> $$x = a\cos(\phi) = r\sin(\theta)\cos(\phi)$$
> $$y = a\sin(\phi) = r\sin(\theta)\sin(\phi)$$
>
> There are a couple of things to note: firstly, we have listed z first, as it is the term which drops out the easiest; secondly, finding the x and y coordinates involves finding the dimension $a = r\sin(\theta)$; this can be thought of as the 'shadow which r casts on the x, y plane'. Once we know this, the x and y coordinates may be found from the angle ϕ.

The value in using polar coordinates rather than a Cartesian coordinate system can be seen in the context of the hydrogen atom pictured in Key Concept 5.4. If we consider electron energies classically, the electron will have two components to consider, namely its kinetic energy and its potential energy. The potential energy, V_e, is a coulombic potential between the electron's charge (q_e) and the proton's charge (q_p), which depends only on the radial separation between two charges (r in our example):

$$V_e = \frac{q_e q_p}{4\pi\varepsilon_0 r} \tag{5.9}$$

In contrast, the kinetic energy can be expressed as a rate of change of the angle around the nucleus. By working in polar coordinates, we can easily separate the coulombic and potential energy terms and consider them independently; something which would not be possible when working in Cartesian coordinates.[‡]

 This is a classical interpretation, now regarded as incorrect. In quantum chemistry, energy considerations are more complex than this, but polar representations still simplify the mathematics.

[‡]The analogy works at a macroscopic level as well; as we move over the Earth's surface (at constant altitude!) our gravitational potential remains constant (constant r from Earth's centre), while our geographical position (longitude and latitude) changes, giving an indication of our kinetic energy over the surface.

5.3.3 Consistency in Converting Polar to Cartesian Coordinates

When we convert from polar coordinates to Cartesian coordinates, we need to ensure that we are correctly preserving our positive and negative components. We can quickly check this with reference to a series of positions, A, B, C and D, shown in Figure 5.14.

For the sake of convenience, we have used a common radius, r. However, consider the four points, A, B, C and D, and what the *sign* of their Cartesian values should be. Our initial concern is, 'Do the conversion factors from Key Concepts 5.5 and 5.6 match up?' When we identify the values that ϕ has for each of the four quadrants and the positive or negative nature of the sines and cosines of each of these, we see that they do indeed match up in two dimensions. The results are given in Table 5.4 and, since the z-coordinate in three dimensions is based on a cosine function, this conversion will also return the correct 'sign' of the coordinate.

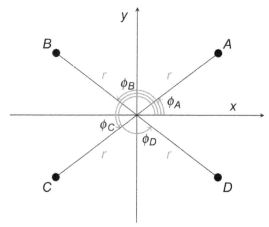

Figure 5.14   Four points defined in Cartesian space and in polar space. All four have constant radius, with each in a different quadrant of the Cartesian space.

Table 5.4   Demonstrating the validity of the polar-to-Cartesian functions in terms of the positive or negative values returned for the x and y coordinates.

Point	Angle range	x coordinate	y coordinate
A	$0 < \phi_A < \dfrac{\pi}{2}$	$r\cos(\phi_A) = \text{positive}$	$r\sin(\phi_A) = \text{positive}$
B	$\dfrac{\pi}{2} < \phi_B < \pi$	$r\cos(\phi_B) = \text{negative}$	$r\sin(\phi_B) = \text{positive}$
C	$\pi < \phi_C < \dfrac{3\pi}{2}$	$r\cos(\phi_C) = \text{negative}$	$r\sin(\phi_C) = \text{negative}$
D	$\dfrac{3\pi}{2} < \phi_D < 2\pi$	$r\cos(\phi_D) = \text{positive}$	$r\sin(\phi_D) = \text{negative}$

> ❶ **Key Concept 5.7 Converting Cartesian to Polar Coordinates**
>
> While the conversion from polar to Cartesian coordinates is fairly straightforward, the reverse is not so simple. Looking at Figure 5.6, we can see that, if $\sin(\phi) = 0.8$, in the range $0:2\pi$, there are two angles that can give this result. The same is true for cosine functions and tangent functions. Therefore, we need to look at our Cartesian values to see in which quadrant our point lies (Figure 5.14) and consider how to process the angle that we calculate from the inverse function.
>
> We convert from Cartesian to polar in two dimensions as:
>
> $$r = \sqrt{x^2 + y^2} \quad ; \quad \phi = \arctan\left(\frac{y}{x}\right)$$
>
> and in three dimensions as:
>
> $$r = \sqrt{x^2 + y^2 + z^2} \quad ; \quad \phi = \arctan\left(\frac{y}{x}\right) \quad ; \quad \theta = \arccos\left(\frac{z}{r}\right)$$
>
> We then apply a correction to ϕ, depending on which quadrant our point lies in.
>
> | Quadrant | x coordinate | y coordinate | Adjustment |
> |---|---|---|---|
> | A | Positive | Positive | Use calculated value. |
> | B | Negative | Positive | Add π to calculated value. |
> | C | Negative | Negative | Add π to calculated value. |
> | D | Positive | Negative | Add 2π to calculated value. |
>
> Note: A correction is not needed for finding θ from the z-coordinate, as $\cos(\theta)$ is single-valued in the range $0:\pi$.

❓Exercise 5.6 Coordinate Systems

For each of these sets of polar coordinates, convert to Cartesian coordinates. Remember that polar coordinates appear in the order (r, ϕ, θ) (three dimensions) or (r, ϕ) (two dimensions).

1. (1.50, 3.62, 1.71)
2. (2.14, 2.89, 2.34)
3. (1.38, 3.47, 2.70)
4. (3.44, 3.52, 0.398)
5. (1.44, 5.65, 1.05)
6. (0.776, 4.43, 2.84)

For each of the sets of Cartesian coordinates below, convert to polar coordinates.

7. (−0.947, −1.35, 1.17)
8. (−0.419, −0.338, 3.68)
9. (−0.927, 0.751, −2.48)
10. (0.403, 0.094, 1.76)

Make sure that your calculator is in radian mode.

Atomic orbitals are a curious example; we use polar coordinates to plot their shapes mathematically, while simultaneously using the Cartesian coordinate system to qualitatively describe their orientations!

5.3.4 *In Context: Atomic Orbital Visualisations*

Atomic orbitals are most easily visualised in spherical polar coordinates rather than three-dimensional Cartesian coordinates. As we have already described, if we think about energies under classical physics, the use of polar systems allows the separation of potential and kinetic energies. This is one way to model the *wave functions* of the electron around a hydrogen atom. In the quantum world, things are more complex; however, polar coordinates still simplify the problem. The image on the cover of this text arises from the polar descriptions of the hydrogen electron wave functions for the 1s, $2p_z$, $3d_{z^2}$, $3d_{yz}$, $4f_{z^3}$ and $4f_{xz^2}$ orbitals. The mathematical forms of these orbitals can be complex; however, they can be described by the following angular dependencies:

1. The 1s orbital has no angular dependence, so appears as a sphere.
2. The $2p_z$, $3d_{z^2}$ and $4f_{z^3}$ orbitals all vary with $(\cos \theta)$, so only show variation along the polar (z) axis.
3. The $3d_{yz}$ and $4f_{xz^2}$ orbitals vary with the product $(\sin \theta \cos \theta \cos \phi)$, showing variation along the polar axis as well as the azimuth axis.

These are gross simplifications of the wave functions of atomic orbitals as it is not the purpose of this text to discuss these wave functions in detail; rather, the intention is to demonstrate how polar coordinates and the trigonometric functions are intrinsically linked in the modelling of chemical processes.

5.4 Summary

In this chapter, we covered the use of trigonometric functions in chemistry; including the definitions of the sine, cosine and tangent functions (and their corresponding inverse functions) and how these are used to define positions across different coordinate systems.

While Cartesian (x, y, z) coordinates are more readily accessible in our experience, as these offer a way to plot correlated experimental data, they are not necessarily the most straightforward way to model chemical systems. Polar coordinates offer us a means to readily separate the angular components of the coordinates from the radial components; from this it becomes possible to carry out modelling of the type required to study the shapes and geometry of atomic orbitals, such as those featured on the cover of this book.

Summary: Chapter 5

Table of Trigonometric Identities

Table 5.5 Table of trigonometric identities.

$$\sin(\theta) = \frac{\text{opposite}}{\text{hypotenuse}} \qquad \cos(\theta) = \frac{\text{adjacent}}{\text{hypotenuse}} \qquad \tan(\theta) = \frac{\text{opposite}}{\text{adjacent}} \equiv \frac{\sin(\theta)}{\cos(\theta)}$$

$$\theta = \arcsin\left(\frac{\text{opposite}}{\text{hypotenuse}}\right) \qquad \theta = \arccos\left(\frac{\text{adjacent}}{\text{hypotenuse}}\right) \qquad \theta = \arctan\left(\frac{\text{opposite}}{\text{adjacent}}\right)$$

$$\operatorname{cosec}(\theta) = \frac{1}{\sin(\theta)} \qquad \sec(\theta) = \frac{1}{\cos(\theta)} \qquad \cot(\theta) = \frac{1}{\tan(\theta)}$$

$$\sin^2(\theta) \equiv (\sin\theta)^2 \qquad \cos^2(\theta) \equiv (\cos\theta)^2 \qquad \tan^2(\theta) \equiv (\tan\theta)^2$$

$$\sin^{-1}(\theta) \equiv \arcsin(\theta) \qquad \cos^{-1}(\theta) \equiv \arccos(\theta) \qquad \tan^{-1}(\theta) \equiv \arctan(\theta)$$

$$(\sin\theta)^{-1} \equiv \operatorname{cosec}\theta \qquad (\cos\theta)^{-1} \equiv \sec\theta \qquad (\tan\theta)^{-1} \equiv \cot\theta$$

$$\sin(2\theta) = \sin(\theta)\cos(\theta) \qquad \cos(2\theta) = \cos^2(\theta) \qquad \tan(2\theta) = \frac{2\tan(\theta)}{1 - \tan^2(\theta)}$$
$$-\sin^2(\theta)$$

$$\sin^2(\theta) + \cos^2(\theta) = 1$$

Conversion Between Coordinate Systems

Note that any arccos or arcsin function has more than one solution, so it is worth checking both approaches to ensure that you are getting the result you expect. For θ in three dimensions, this is less of an issue, as this only varies from 0 to π.

Table 5.6 Conversion from polar to Cartesian coordinates.

Two dimensions	Three dimensions
$x = r\cos(\phi)$	$z = r\cos(\theta)$
$y = r\sin(\phi)$	$x = r\sin(\theta)\cos(\phi)$
	$y = r\sin(\theta)\sin(\phi)$

Table 5.7 Conversion from Cartesian to polar coordinates.

Two dimensions	Three dimensions
$r = \sqrt{x^2 + y^2}$	$r = \sqrt{x^2 + y^2 + z^2}$
$\phi = \arcsin\left(\dfrac{y}{r}\right) = \arccos\left(\dfrac{x}{r}\right)$	$\theta = \arccos\left(\dfrac{z}{r}\right)$
	$\phi = \arccos\left(\dfrac{x}{r\sin\theta}\right) = \arcsin\left(\dfrac{y}{r\sin\theta}\right)$

?Exercise 5.7 Trigonometry in a Chemical Context

---

1. The Bragg equation, $n\lambda = 2d\sin\theta$, commonly used in crystallography, relates the diffraction angle for X-rays of wavelength λ as a function of the distance between crystal planes, d. The value n is an integer value related to the layered nature of the crystal. An X-ray of wavelength 232 pm is diffracted from a sodium chloride crystal. Given that the separation between the crystal planes is 2.84 Å, calculate the angle at which the first-order $(n=1)$ diffraction peak will be observed.

2. In a similar experiment, the same X-ray shows a first-order diffraction peak at an angle of 35.1°. Calculate the spacing between crystal planes.

3. A two-dimensional layer of a sulfur crystal is based on a repeating parallelogram of side lengths $b = 10.855(2)$ Å and $c = 10.790(3)$ Å. The angle between these sides is $\beta = 95.92(2)°$. Calculate the length of the diagonal across this repeating unit, giving consideration to the uncertainties in the measurement (Chapter 3).[§]

4. A hydrogen atom can be visualised on a Cartesian coordinate system with the proton at the origin. At the instant the electron passes through the point (0.244, 0.128, 0.801) Å, determine the distance of the electron from the nucleus and use this to calculate the coulombic potential *via*

$$V_{coulombic} = \frac{q_{nucleus} \times q_{electron}}{4\pi\varepsilon_0 r}$$

where $q_{nucleus} = -q_{electron} = 1.602 \times 10^{-19}$ C, and $\varepsilon_0 = 8.854 \times 10^{-12}$ J^{-1} C^2 m^{-1}.

5. The carbon dioxide molecule depicted in Figure 5.13 is modelled on a polar coordinate system. Assuming that the carbon atom sits at the origin of the system, determine the Cartesian coordinates for each oxygen atom when one oxygen atom sits at $\theta = 2^c$. The C=O bond length is 116 pm.

[§]Data from L. K. Templeton, D. H. Templeton and A. Zalkin, *Inorg. Chem.*, 1976, **15**(8), 1999–2001. DOI: 10.1021/ic50162a059.

Solutions to Exercises

Solutions: Exercise 5.1

1. 9.29^c
2. 3.26^c
3. 1.3^c
4. 10.9^c
5. 3.05^c
6. 4.21^c
7. $10°$
8. $205°$
9. $59.0°$
10. $446°$
11. $206°$
12. $218°$

Solutions: Exercise 5.2

1. 0.11 cm
2. 1.40 cm
3. 2.14 cm
4. 0.48 cm
5. 3.37 cm
6. 1.6 cm
7. 5.86 cm
8. 0.596 cm

Solutions: Exercise 5.3

1. $\theta = 27°$
2. $\theta = 17°$
3. $\theta = 36°$
4. $\theta = 34°$
5. $\theta = 27°$
6. $\theta = 40°$
7. $\theta = 25°$
8. $\theta = 82°$

Solutions: Exercise 5.4

1. $c = 1.38$ cm
2. $a = 3.28$ cm
3. $b = 3.70$ cm
4. $c = 1.78$ cm or 1.41 cm
5. $b = 3.63$ cm
6. $\beta = 24°$
7. $c = 1.94$ cm
8. $\gamma = 63°$
9. $\gamma = 59°$
10. $\beta = 80°$
11. $c = 5.08$ cm
12. $d_{C-H} = 2.00$ Å

For Questions 4 and 5 the rearrangement of cosine rule is a quadratic which can sometimes have two positive solutions.

Solutions: Exercise 5.5

1. $r = 1.03$ Å
2. $r = 1.23$ Å
3. $r = 1.03$ Å

Solutions: Exercise 5.6

1. (−1.32, −0.684, −0.208)
2. (−1.49, 0.383, −1.49)
3. (−0.558, −0.190, −1.25)
4. (−1.24, −0.493, 3.17)
5. (1.01, −0.739, 0.717)

6. (−0.0642, −0.221, −0.741)
7. (2.02, 4.10, 0.954)
8. (3.72, 3.82, 0.145)
9. (2.75, 2.46, 2.69)
10. (1.81, 0.229, 0.231)

Solutions: Exercise 5.7

1. 24.1°

2. 2.02 Å

3. 16.075 3 ± 5 ≡ 16(5) Å.

4. r = 0.847 Å, 2.72 × 10^{-18} J.

5. Atom 1: (−48.3, 105) pm
   Atom 2: (48.3, −105) pm

Learning Points: What We'll Cover

□ Vectors as quantities having a magnitude and a direction and how to appropriately notate them
□ How to scale and add vectors in two or more dimensions, as well as how to quantify unit vectors
□ Identifying vector quantities in chemistry, with particular focus on positions, fields, dipole moments and forces
□ Vector multiplication, including the 'dot' (scalar) product and the 'cross' (vector) product, and what these mean in the context of chemistry

Why This Chapter Is Important

- Many quantities in chemistry can be represented as having a direction as well as a magnitude, making them vectors.
- Being able to do calculations on vectors allows us to solve problems occurring in more than one dimension, such as the effect of dipole moments in spectroscopy.
- Thinking of quantities as vectors can help visualise and quantify chemical processes, such as the movement of charged species through an electric field.

Vectors, Directions and Crystal Structures

'*What do you get when you cross a mountaineer and a mosquito?*'[†]
This is a joke well recognised by mathematicians, and by the end
of the chapter you will understand the punchline! Vectors appear
everywhere, whether you are a navigator on board an aeroplane, a
footballer shooting for goal on a windy day, a snooker player considering the impacts leading to the perfect pot shot … the list goes
on. In chemistry, we use vectors for a number of different applications; they are used when considering a dipole moment, when
looking at electronic transitions or in the consideration of crystal
structures and unit cells. When we start to consider transfer of
energy, the ability to multiply vector quantities becomes useful as
it can greatly simplify our calculations of chemical models. In this
chapter, we discuss the concept of a vector as a 'directed number',
how we can treat the scalar value of a vector (its size) together with
its direction and how to apply mathematics with these quantities
in order to unlock our understanding of these *directed values*.

6.1 A Brief Introduction to the Vector

A vector in the context of mathematics is a quantity which combines
both magnitude and a direction. An example in the physical world
is velocity; it combines the magnitude of speed together with a
direction. For example, a car's speed may be '30 m s^{-1}'; but its velocity is '30 m s^{-1} due west'. This contrasts with *scalar* quantities which
only have a magnitude associated with them.

Vectors can also be operated on in the same manner as scalar quantities (addition, subtraction, multiplication); however, the rules by
which these must be performed are slightly different.

[†]Nothing; you can't cross a scalar and a vector….

In chemistry, the dipole moment in a molecule may be considered a vector. It has a magnitude (the charge difference between the two ends) and a direction (pointing positive to negative). The 'direction' in the context of 'pointing positive to negative' is straightforward if there are two atoms (the vector will point along the direction of the bond between the two); however, it becomes slightly more troublesome when we start to consider polyatomic molecules. In this situation, we need to have a reference point (an *origin*), as all bonds (and associated dipoles) will be pointing in different directions.

The way in which we notate a vector is significant, and there are a number of ways to notate vectors. This is summarised in Key Concept 6.1; in this text we will be using the bold-font notation in line with many other mathematics texts, *i.e.* a vector is represented as **a**.

The 'hat' (caret) above a vector always denotes a **unit vector** of magnitude 1.

---

❶ Key Concept 6.1 Vector Notation

Vectors are always notated with letters, and in a **bold font** in printed work. They can also be indicated in other ways to show particular concepts. Vectors are also commonly underlined where bold is not available (*e.g.* handwriting!).

Notation	Meaning	Notes
$\mathbf{a}, \bar{a}, \underline{a}$	Vector, **a**	Equivalent representations of a vector; bold font tends to be used in printed text, while underlines tend to be used in handwritten examples.
$\lvert\mathbf{a}\rvert, \lvert\bar{a}\rvert$ or $\lvert\underline{a}\rvert$	Modulus of **a**	Reports only the *magnitude* of vector **a**.
$\hat{\mathbf{a}}$	Unit vector of **a**	A vector in the *direction* of **a**, but with magnitude 1.
$\hat{\mathbf{i}}, \hat{\mathbf{j}}$ or $\hat{\mathbf{k}}$	Standard unit vectors	Vectors of magnitude 1 in the mutually perpendicular i, j and k directions.
$\begin{pmatrix} a \\ 0 \\ c \end{pmatrix}$	$a\hat{\mathbf{i}} + 0\hat{\mathbf{j}} + c\hat{\mathbf{k}}$	Column, or *matrix* notation. Can make vector algebra easier, but must always include all three dimensions, even in a two-dimensional problem.

> ❗ **Key Concept 6.2 Unit Vectors**
>
> A unit vector is simply a vector with unit magnitude (*i.e.* magnitude of 1). Its sole purpose is to indicate a direction; it is then scaled by whatever value we wish.
>
> A vector can then be represented in terms of its unit vector:
>
> $$\mathbf{c} = 6\hat{\mathbf{c}}$$
>
> $$\mathbf{d} = 0.4\hat{\mathbf{d}}$$
>
> In this case, the *direction information* is held in the unit vector, while the number of steps taken (the *scalar*) is simply the multiple of this unit vector.
>
> The standard unit vectors $\hat{\mathbf{i}}$, $\hat{\mathbf{j}}$ and $\hat{\mathbf{k}}$ may be considered to be unit vectors running along the Cartesian *x*-, *y*- and *z*-axes, respectively.

6.1.1 *Vector Addition in One Dimension*

A one-dimensional vector acts along a single axis. In this dimension, it can have two directions: either in the direction of the axis, or in the opposite direction, against the axis. This is handled simply by using either a 'plus' (+) or a 'minus' (−) to indicate positive or negative directions. If we walk in a straight line, we can picture this as, 'Take two steps forward,' and then, 'Take one step back.' We can see that if we translate this into mathematics, our overall progress will be $2 + (-1) = 1$ step forward.

We use the same principle in concepts in areas such as thermodynamics; the notation of 'change' (*e.g.* $\Delta_{fus}H$) requires a 'direction' so that we can appropriately use it in Hess's law; the direction of change tells us whether energy flows 'out' of the system (negative) or whether energy flows 'into' the system (positive).

We can take this principle of combining positive and negative vectors in a single dimension and now apply it when thinking about higher-dimensional vectors.

6.2 Two (and Higher) Dimensional Vectors

In two dimensions, we can no longer use 'positive' and 'negative' numbers alone to describe the direction of the vector. Instead, we use a system analogous to the *Cartesian coordinate system*, discussed in Chapter 5, with which we are already familiar.

If we consider the example of steps introduced in one dimension (Section 6.1.1), we can extend this to two dimensions:

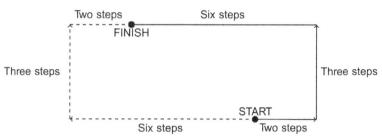

Figure 6.1   Showing the effect of taking steps in different orders. It doesn't matter which order we take the steps in as we will always end up in the same place.

Two steps to the right, three steps forward, six to the left...

Looking at this example, we see that we have the elements of a coordinate system; in effect, we have created a list of instructions. Let's firstly define our 'left–right' motion to be along the x-axis, and our 'forward–back' motion along the y-axis. Our instructions then become:

1. Take two steps in the $+x$ direction.
2. Take three steps in the $+y$ direction.
3. Take six steps in the $-x$ direction.

We have numbered the steps; however, there is a question over whether it matters what order we carry these steps out. This is shown in Figure 6.1, demonstrating that it doesn't matter what order the steps are taken, as we will still end up in the same location relative to the start point.

The $\hat{\mathbf{i}}, \hat{\mathbf{j}}$ and $\hat{\mathbf{k}}$ unit vectors are *orthogonal* (Key Concept 6.3); this means that it is impossible to progress in the i direction by only making steps in the j direction and so on.

What we have done is defined vectors as *components*; we have a 'forward–back' component and a 'left–right' component, neither of which affects the other. Should we wish to move beyond two dimensions, we can introduce a third component: 'up–down'. Mathematically, this defines an (x, y, z) coordinate system, and we extend our principle of unit vectors accordingly. By convention, the unit vectors along the x-, y- and z-axes are labelled $\hat{\mathbf{i}}, \hat{\mathbf{j}}$ and $\hat{\mathbf{k}}$, respectively. These are introduced in Key Concepts 6.1 and 6.2, and these letters should be avoided unless explicitly referring to these unit vectors. By expressing two- and three-dimensional vectors as a series of orthogonal unit vectors (see Key Concept 6.3), we create a set of building blocks which we can use to perform vector algebra.

> **❶ Key Concept 6.3 Orthogonality: The Independence of Unit Vectors**
>
> The term *orthogonal* sounds more complicated than it actually is. It is analogous to the term *perpendicular*; however, it applies to all dimensions, not simply our three-dimensional world.
>
> Two quantities are considered to be *orthogonal* if they are completely independent of one another. In vectors, we consider this to mean that the cosine of the angle between the vectors is equal to zero.
>
> In *Cartesian coordinates*, it is easy to see that the x-, y- and z-axes are mutually perpendicular (and are also orthogonal); however, when we consider *polar coordinates* (see Section 5.3.2), the radius and angular measurements are not perpendicular, but are independent and, as such, are deemed to be orthogonal.

6.3 Vector Addition, Subtraction and Scaling

When we write vectors in terms of their components $\hat{\mathbf{i}}$, $\hat{\mathbf{j}}$ and $\hat{\mathbf{k}}$, combining them is simply a matter of adding together all the $\hat{\mathbf{i}}$ components, all the $\hat{\mathbf{j}}$ components and all the $\hat{\mathbf{k}}$ components. This process of combining vectors is called finding a *resultant* vector; a single vector which encompasses all the steps in all contributing vectors. This is shown in Example 6.1, in which vectors are added in three-dimensional space.

This approach can be used to determine the direction of the dipole moments (Example 6.2) in a molecule, which can aid in the prediction of its spectroscopic properties[‡] and calculations of its centre of mass and *moment of inertia*, essential for rotational spectroscopy.

> 💬 The moment of inertia can be thought of as a molecule's ability to maintain rotational kinetic energy; the greater its moment of inertia, the more rotational kinetic energy it can store for a given rate of rotation.

Example 6.1 Finding a Resultant Vector

Q: Two vectors, **a** and **b**, are described by the unit vectors:

$$\mathbf{a} = 3\hat{\mathbf{i}} + 5\hat{\mathbf{j}} + \hat{\mathbf{k}}$$

$$\mathbf{b} = 6\hat{\mathbf{i}} - \hat{\mathbf{j}}$$

Find the resultant vector, **c**.

[‡]While this approach gives us information on the direction of our dipole moments, it cannot directly predict the absolute dipole moment of a molecule *i.e.* the scalar value of the unit vector. More information is needed in order to make these predictions, however this is outwith the scope of this text.

A: We can write **c** as a sum of the two vectors, as:

$$\mathbf{c} = \mathbf{a} + \mathbf{b}$$

$$= \left(3\hat{\mathbf{i}} + 5\hat{\mathbf{j}} + \hat{\mathbf{k}}\right) + \left(6\hat{\mathbf{i}} - \hat{\mathbf{j}}\right)$$

We can now consider the $\hat{\mathbf{i}}, \hat{\mathbf{j}}$ and $\hat{\mathbf{k}}$ components separately, adding the two $\hat{\mathbf{i}}$ terms together, then the two $\hat{\mathbf{j}}$ terms together and then the two $\hat{\mathbf{k}}$ terms together to find our final resultant vector.

$$\mathbf{c} = [3 + 6]\hat{\mathbf{i}} + \left[5 + (-1)\right]\hat{\mathbf{j}} + [1 + 0]\hat{\mathbf{k}}$$

$$= 9\hat{\mathbf{i}} + 4\hat{\mathbf{j}} + \hat{\mathbf{k}}$$

It might look as if there is only one $\hat{\mathbf{k}}$, but this is because we should consider the vector **b** to be written as $\mathbf{b} = 6\hat{\mathbf{i}} - 1\hat{\mathbf{j}} + 0\hat{\mathbf{k}}$

When considering dipole moments in a simple molecule, such as CO_2, it should be fairly clear from inspection that the overall molecular dipole moment is zero. Since the molecule is linear, the two dipoles running along each bond cancel each other out exactly. However, in molecules with a lower-symmetry structure, such as water, we can use vectors to determine the overall effect of each contributing bond dipole to the overall molecular dipole moment.

To determine these vectors, we need to know the position of each of the atoms in a molecule; each of the atoms can be represented by a position vector, identifying its unique location in three-dimensional space. These position vectors can then be used to determine the direction of the dipole moment for each bond. A position vector is denoted **r**, with each atom carrying a subscript identifying the atom—so for water (Example 6.2), example notations, such as $\mathbf{r}_O$, $\mathbf{r}_{H_a}$ and $\mathbf{r}_{H_b}$ or $\mathbf{r}_O$, $\mathbf{r}_{H_1}$ and $\mathbf{r}_{H_2}$ are both valid as both clearly differentiate between the two different hydrogen atoms.

The coordinates are presented as (i, j); a value for k is not listed, as water exists in a two-dimensional plane and so only two dimensions (i and j) are needed to fully describe the location of all the atoms.

Recall that, by convention, a *position vector* carries the symbol **r**.

Example 6.2 Finding a Resultant Vector in a Chemical Context

Q: A water molecule is oriented on a two-dimensional plane such that the oxygen atom lies at the origin and the hydrogen atoms lie at (0.96, 0) and (−0.32, 0.94). Given that the bond dipoles point from the hydrogen atom to the oxygen, calculate the direction of the vector for the resultant dipole moment. A sketch may help.

A: This problem is much the same as Example 6.1, and we will work through it to find the direction of the resultant vector.

Firstly, identify the position vectors for each of the atoms. We create these from the provided Cartesian coordinates of each of the atoms:

$$\mathbf{r}_O = 0\hat{\mathbf{i}} + 0\hat{\mathbf{j}}$$

$$\mathbf{r}_{H_a} = 0.96\hat{\mathbf{i}} + 0\hat{\mathbf{j}}$$

$$\mathbf{r}_{H_b} = -0.32\hat{\mathbf{i}} + 0.94\hat{\mathbf{j}}$$

Now, we have been told that the dipole vector points from the hydrogen to the oxygen. Therefore, the dipole vector can be found by subtracting the starting point (the hydrogen) from the finish point (the oxygen). We will use the symbol μ to denote the dipole moment vector.

$$\mu_a = \mathbf{r}_0 - \mathbf{r}_{H_a}$$
$$= -0.96\hat{\mathbf{i}} + 0\hat{\mathbf{j}}$$
$$\mu_b = \mathbf{r}_0 - \mathbf{r}_{H_b}$$
$$= +0.32\hat{\mathbf{i}} - 0.94\hat{\mathbf{j}}$$

Now that we have the direction vectors for each of our dipoles, we simply need to add them together to get the direction of the dipole moment for the molecule:

$$\mu_{H_2O} = \mu_a + \mu_b$$
$$= \left(-0.96\hat{\mathbf{i}} + 0\hat{\mathbf{j}}\right) + \left(0.32\hat{\mathbf{i}} - 0.94\hat{\mathbf{j}}\right)$$
$$= (-0.96 + 0.32)\hat{\mathbf{i}} + (0 - 0.94)\hat{\mathbf{j}}$$
$$= -0.64\hat{\mathbf{i}} - 0.94\hat{\mathbf{j}}$$

This now tells us the direction of the dipole moment within water. This result may be combined with information about the electronegativities of the atoms involved to determine the 'size' of this dipole. In water, because both bonds are equivalent (O–H), each will have the same dipole moment; however, when there are a number of different types of bond to consider, the 'weighting' of the dipole moment (based on the difference of electronegativities of the two atoms) must be considered.

In chemistry, change is *always* 'final − initial', e.g. ΔG is defined as $G_{final} - G_{initial}$.

In computational chemistry, position vectors and atom electronegativities are used to determine overall molecular properties, such as the overall molecular dipole moment. Although we rarely calculate these properties by hand, it is useful to understand the mathematics of the computer simulations to understand the results.

6.3.1 Vector Scaling

A vector can be scaled in the same way as an algebraic quantity can be scaled, and each of its components, $\hat{\mathbf{i}}$, $\hat{\mathbf{j}}$ and $\hat{\mathbf{k}}$, is scaled correspondingly, as:

$$\mathbf{a} = 3\hat{\mathbf{i}} + 4\hat{\mathbf{j}}$$
$$5\mathbf{a} = 5\left(3\hat{\mathbf{i}} + 4\hat{\mathbf{j}}\right)$$
$$= 15\hat{\mathbf{i}} + 20\hat{\mathbf{j}} \tag{6.1}$$

The consequence of this is that the position vector $3\mathbf{r}$ will be three times farther away from the origin than $\mathbf{r}$ but in the same direction.

6.3.2   *In Context: Finding the Centre of Mass*

An understanding of how vectors can be represented, managed and combined can greatly simplify problems in three dimensions. If we know the *position vector* of all atoms in a molecule, we can determine the position of the centre of mass (and hence the centre of rotation) by finding the 'mean' position vector, as shown in eqn (6.2). Here, the sum of all the 'moments' (mass, m, times position vector) is divided by the total mass of the molecule, M_{tot}:

$$\mathbf{r}_{centre} = \frac{1}{M_{tot}}\left[m_1\mathbf{r}_1 + m_2\mathbf{r}_2 + \ldots + m_n\mathbf{r}_n\right] \qquad (6.2)$$

Example 6.3 Determining a Centre of Mass

Q: The molecule dichloromethane (CH_2Cl_2, Figure 6.2) has its atoms at the following position vectors:

$$\mathbf{r}_C = \left(0\hat{\mathbf{i}} + 0\hat{\mathbf{j}} + 0\hat{\mathbf{k}}\right)Å$$

$$\mathbf{r}_{H_a} = \left(-0.89\hat{\mathbf{i}} + 0\hat{\mathbf{j}} + 0.60\hat{\mathbf{k}}\right)Å$$

$$\mathbf{r}_{H_b} = \left(0.89\hat{\mathbf{i}} + 0\hat{\mathbf{j}} + 0.60\hat{\mathbf{k}}\right)Å$$

$$\mathbf{r}_{Cl_a} = \left(0\hat{\mathbf{i}} + 1.47\hat{\mathbf{j}} - 0.99\hat{\mathbf{k}}\right)Å$$

$$\mathbf{r}_{Cl_b} = \left(0\hat{\mathbf{i}} - 1.47\hat{\mathbf{j}} - 0.99\hat{\mathbf{k}}\right)Å$$

Calculate the position of its centre of mass, assuming ^1H and ^{35}Cl.

A: Using eqn (6.2), we multiply the position vectors by the atomic masses to obtain a 'weighted position vector'. We then add all of these vectors together and divide by the overall molecular mass of CH_2Cl_2 ($M_{tot} = 84$ Da) to find the position of the centre of mass.

$$m_C\mathbf{r}_C = 12\,Da\left(0\hat{\mathbf{i}} + 0\hat{\mathbf{j}} + 0\hat{\mathbf{k}}\right)Å \qquad = \left(0\hat{\mathbf{i}} + 0\hat{\mathbf{j}} + 0\hat{\mathbf{k}}\right)Å\,Da$$

$$m_H\mathbf{r}_{H_a} = 1\,Da\left(-0.89\hat{\mathbf{i}} + 0\hat{\mathbf{j}} + 0.60\hat{\mathbf{k}}\right)Å = \left(-0.89\hat{\mathbf{i}} + 0\hat{\mathbf{j}} + 0.60\hat{\mathbf{k}}\right)Å\,Da$$

$$m_H\mathbf{r}_{H_b} = 1\,Da\left(0.89\hat{\mathbf{i}} + 0\hat{\mathbf{j}} + 0.60\hat{\mathbf{k}}\right)Å \quad = \left(0.89\hat{\mathbf{i}} + 0\hat{\mathbf{j}} + 0.60\hat{\mathbf{k}}\right)Å\,Da$$

$$m_{Cl}\mathbf{r}_{Cl_a} = 35\,Da\left(0\hat{\mathbf{i}} + 1.47\hat{\mathbf{j}} - 0.99\hat{\mathbf{k}}\right)Å = \left(0\hat{\mathbf{i}} + 51.45\hat{\mathbf{j}} - 34.65\hat{\mathbf{k}}\right)Å\,Da$$

$$m_{Cl}\mathbf{r}_{Cl_b} = 35\,Da\left(0\hat{\mathbf{i}} - 1.47\hat{\mathbf{j}} - 0.99\hat{\mathbf{k}}\right)Å = \left(0\hat{\mathbf{i}} - 51.45\hat{\mathbf{j}} - 34.65\hat{\mathbf{k}}\right)Å\,Da$$

$$\text{Weighted total} = \left(0\hat{\mathbf{i}} + 0\hat{\mathbf{j}} - 1\hat{\mathbf{k}}\right)Å\,Da$$

$$\frac{\text{Weighted total}}{M_{tot}} = \mathbf{r}_{center} = \left(0\hat{\mathbf{i}} + 0\hat{\mathbf{j}} - 0.81\hat{\mathbf{k}}\right)Å$$

Interpreting this result, if the molecule is oriented such that the chlorine atoms are below the carbon atoms, the centre of mass, O, lies 0.81 Å directly below the carbon atom.

It is usually enough to approximate the 'mass of an isotope' as its nucleon number but this *is* an approximation.

Da is the unit dalton—it is 1/12th the mass of a ^{12}C atom.

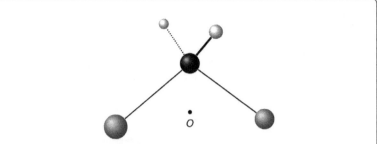

Figure 6.2   Dichloromethane, CH_2Cl_2, with the centre of mass, O, marked; the centre of mass lies in the same plane as the Cl–C–Cl plane.

The centre of mass is the point around which the molecule will rotate, so knowing the position of its centre of mass allows calculation of the moment of inertia for rotational spectroscopy.

? Exercise 6.1 Vector Addition

For the following vectors:

$\mathbf{a} = 6\hat{\mathbf{i}} + 7\hat{\mathbf{j}}$ $\mathbf{d} = 3\hat{\mathbf{i}} - 8\hat{\mathbf{j}} - 6\hat{\mathbf{k}}$

$\mathbf{b} = 7\hat{\mathbf{i}} - 5\hat{\mathbf{j}}$ $\mathbf{f} = 9\hat{\mathbf{i}} + 7\hat{\mathbf{j}} - 2\hat{\mathbf{k}}$

$\mathbf{c} = -3\hat{\mathbf{j}} + 7\hat{\mathbf{k}}$ $\mathbf{g} = -8\hat{\mathbf{i}} - 3\hat{\mathbf{j}} + 5\hat{\mathbf{k}}$

Evaluate the following vector summations:

1. $\mathbf{a} + \mathbf{b}$ 5. $\mathbf{c} + \mathbf{f}$
2. $\mathbf{b} - \mathbf{c}$ 6. $\mathbf{f} - \mathbf{b}$
3. $\mathbf{d} - \mathbf{f}$ 7. $3\mathbf{a} + \mathbf{f}$
4. $\mathbf{g} + \mathbf{d}$ 8. $\mathbf{g} - 4\mathbf{b}$

We will use these same combinations in later exercises to demonstrate the different results possible, depending on how we combine the vectors.

6.4 Vector Multiplication: Work Done and Torque

We have seen how to add and subtract vectors through consideration of their components. The logical question now is how to move to higher-order operations, *i.e.* multiplication. There are some examples in chemistry in which it is useful to consider how vectors can multiply together. The first example is in calculating the 'work done' as an object is moved over a distance; whether

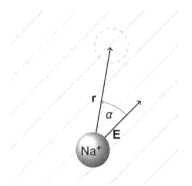 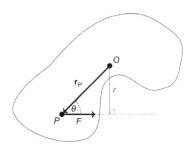

(a) A sodium ion, Na⁺, moves through an electric field **E** to a new position **r**; the work done is found from the product of the electric field and the distance moved in the direction of the electric field.

(b) A force **F** acts at a specific point *P*; its torque around the centre of rotation *O* is found from the perpendicular distance *r* of the force's line of action.

Figure 6.3   Examples of vectors in context; an ion migrating through an electric field and a torque applied to a rotating body.

this is an ion moving through an electric field (Figure 6.3a) or an adsorbed molecule migrating over a surface. Either way, this is calculated as 'force times distance' but it can only be easily calculated if the force is perfectly aligned with the direction of movement. By considering both force and distance as vectors, we can calculate work done no matter the alignment of the force or velocity vectors.

Similarly, a torque (or 'twisting force') around a pivot is only calculated easily when we know the perpendicular distance of the force's 'line of action' from the centre of rotation (Figure 6.3b). This has applications in rotational spectroscopy of molecules. Again, using vector multiplication can greatly simplify this exercise.

To start our exploration of vector multiplication, it would be sensible to take what appears to be the logical step: to write our vectors as 'bracketed quantities' and multiply them together. Let's consider our two vectors; **a** and **b**:

$$\mathbf{a} = 3\hat{\mathbf{i}} + 4\hat{\mathbf{j}}$$
$$\mathbf{b} = 5\hat{\mathbf{i}} - 6\hat{\mathbf{j}} \tag{6.3}$$

If we take these vectors and multiply them together in the same manner as an expanded bracket, we get an unusual result, blending multiplied unit vectors and multiplied scalar values (eqn (6.4); see Toolkit A.2 if you are unsure about what is happening):

$$\mathbf{a} \times \mathbf{b} = \left(3\hat{\mathbf{i}} + 4\hat{\mathbf{j}}\right) \times \left(5\hat{\mathbf{i}} - 6\hat{\mathbf{j}}\right)$$
$$= 15\left(\hat{\mathbf{i}} \times \hat{\mathbf{i}}\right) - 24\left(\hat{\mathbf{j}} \times \hat{\mathbf{j}}\right) - 18\left(\hat{\mathbf{i}} \times \hat{\mathbf{j}}\right) + 20\left(\hat{\mathbf{j}} \times \hat{\mathbf{i}}\right) \tag{6.4}$$

This looks like a mess, not least of which because the result appears to makes no sense! For reasons which will become apparent later, *the order in which we multiply our vectors matters!* We must answer some important questions, however; namely, 'What happens when we multiply perpendicular vectors?' (*e.g.* $\hat{\mathbf{i}} \times \hat{\mathbf{j}}$) and 'What happens when we multiply parallel vectors?' (*e.g.* $\hat{\mathbf{i}} \times \hat{\mathbf{i}}$).

As it turns out, we get two different results, depending on how we answer these questions and, because of this, there are two ways to multiply vectors together. The first of these is called the *vector dot product* (represented by the 'dot' multiplication notation, $\mathbf{a}\cdot\mathbf{b}$), while the second is called the *vector cross product* (represented by the 'cross' multiplication notation, $\mathbf{a}\times\mathbf{b}$). From this point on in the chapter, it is vital to pay attention to whether we are using a 'dot' or a 'cross' to indicate the multiplication!

> When multiplying scalar quantities, the 'times sign' $\times$ and the 'dot sign' $\cdot$ are often used interchangeably, but in vector multiplication they have two different meanings.

6.4.1 The 'Dot' (Scalar) Product

The vector dot product is also called the *scalar product* because the result is a scalar value with no direction. The definition of the dot product is shown in Key Concept 6.4. This can be found from the magnitudes of the vectors involved and the cosine of the angle between them. The consequence of this is that vectors which are perpendicular to each other (such as the $\hat{\mathbf{i}}$, $\hat{\mathbf{j}}$ and $\hat{\mathbf{k}}$ unit vectors) are cancelled out (as $\cos(\pi/2) = 0$), while vectors which are parallel to each other (such as two of either of the previous unit vectors, *e.g.* $\hat{\mathbf{i}}$ and $\hat{\mathbf{i}}$) simply multiply together (as $\cos(0) = 1$).

❗ Key Concept 6.4 The Vector Dot Product

The definition of the dot product is:

$$\mathbf{m}\cdot\mathbf{p} = |\mathbf{m}||\mathbf{p}|\cos\theta$$

where θ is the angle between the vectors. This has the effect that parallel vectors are directly multiplied together (since $\theta = 0$ and $\cos(0) = 1$), while perpendicular vectors $\theta = \pi/2$ are cancelled out $\cos(\pi/2) = 0$:

$$3\hat{\mathbf{i}} \cdot 4\hat{\mathbf{i}} = 12\left(\hat{\mathbf{i}} \cdot \hat{\mathbf{i}}\right) = 12 \cdot (1) = 12$$

$$3\hat{\mathbf{i}} \cdot 5\hat{\mathbf{j}} = 15\left(\hat{\mathbf{i}} \cdot \hat{\mathbf{j}}\right) = 15 \cdot (0) = 0$$

> Remember that an angle of $\pi/2^c$ is equivalent to an angle of $90°$. See Chapter 5 for discussion on the use of radians.

The definition of the dot product gives us a useful tool to find the angle between any two vectors. This is shown in Example 6.4, while we demonstrate an application within chemistry in Example 6.5.

Example 6.4 Determining the Angle Between Vectors Using the Dot Product

Consider our earlier pair of vectors, **a** and **b**, and their dot product:

$$\mathbf{a} \cdot \mathbf{b} = \left(3\hat{\mathbf{i}} + 4\hat{\mathbf{j}}\right) \cdot \left(5\hat{\mathbf{i}} - 6\hat{\mathbf{j}}\right)$$

$$= 15\left(\hat{\mathbf{i}} \cdot \hat{\mathbf{i}}\right) + (-24)(\hat{\mathbf{j}} \cdot \hat{\mathbf{j}}) + (-18)\left(\hat{\mathbf{i}} \cdot \hat{\mathbf{j}}\right) + 20\left(\hat{\mathbf{j}} \cdot \hat{\mathbf{i}}\right)$$

Remember that parallel unit vectors 'dot' to 1, while perpendicular unit vectors 'dot' to 0:

$$\mathbf{a} \cdot \mathbf{b} = 15(1) - 24(1) - 18(0) + 20(0)$$

$$= -9$$

Now that we have a value for our 'dot' product, we need to calculate the magnitude (the 'modulus') of our vectors, **a** and **b**. We can again apply Pythagoras' theorem:

$$|\mathbf{a}| = \sqrt{3^2 + 4^2} \quad |\mathbf{b}| = \sqrt{5^2 + 6^2}$$

$$= \sqrt{25} \qquad\qquad = \sqrt{61}$$

$$= 5 \qquad\qquad\quad = 7.81$$

We can now use the definition of the dot product given in Key Concept 6.4:

$$\mathbf{a} \cdot \mathbf{b} = |\mathbf{a}||\mathbf{b}| \cos\theta$$

Rearranging for $\cos\theta$ gives:

$$\cos\theta = \frac{\mathbf{a} \cdot \mathbf{b}}{|\mathbf{a}||\mathbf{b}|}$$

$$= \frac{-9}{39.05}$$

This gives $\cos\theta = -0.231$, representing an angle of $103°$ (1.80^c) between the vectors.

Electric fields can be reported as a force per unit charge, or a potential difference per unit distance. These two units are congruent, and either can be used, but here 'force per unit charge' (with the unit newton per coulomb) will be more useful to us.

Example 6.5 Determination of Electrical Work

Q: A sodium ion, Na^+ is moved across a uniform electric field from position $\mathbf{r}_1 = \left(\hat{\mathbf{i}} - 3\hat{\mathbf{j}}\right)$ cm to $\mathbf{r}_2 = \left(-2\hat{\mathbf{i}} + 2\hat{\mathbf{j}}\right)$ cm. Given that the electric field vector is $\mathbf{E} = \left(\hat{\mathbf{i}} + 5\hat{\mathbf{j}}\right)$ N C^{-1}, calculate the electrical work, W_e transferred and determine whether it is work done on the particle or work done by the particle. The work done is found from $\mathbf{F} \cdot \mathbf{r}$ (a visual representation of this is shown in Figure 6.3a).

A: Firstly, we need to identify the vector defining the motion of Na^+; recall that this is a 'final − initial' situation, so we take the position vector of the end point and subtract the starting point from this, remembering to convert to SI units:

$$r = r_2 - r_1$$
$$= \left(-2\hat{i} + 2\hat{j}\right)cm - \left(\hat{i} - 3\hat{j}\right)cm$$
$$= \left(-3\hat{i} + 5\hat{j}\right)cm$$
$$= \left(-3\hat{i} + 5\hat{j}\right) \times 10^{-2}m$$

We now need to determine the force vector acting on the particle. Given that the unit of the electric field is N C^{-1}, we can calculate the force vector by multiplying the field by the charge on the sodium ion ($q = +1.602 \times 10^{-19}$ C). As q is a simple scalar value, we will apply its value towards the end of our calculation:

$$\mathbf{F}_{elec} = q\mathbf{E}$$

We now have all of the information needed to solve the problem.

$$W_e = \mathbf{F} \cdot \mathbf{r} = q(\mathbf{E} \cdot \mathbf{r})$$
$$= q\left[\left(\hat{i} + 5\hat{j}\right) \cdot \left(-3\hat{i} + 5\hat{j}\right) \times 10^{-2}m\right]$$
$$= q[-3 + 25] \times 10^{-2}m = (22q) \times 10^{-2}m \ NC^{-1}$$

This gives our final result as 22 × 10^{-2} m × 1.602 × 10^{-19} C = 3.524 × 10^{-20} J.

Remember: $\hat{i} \cdot \hat{i} = 1$ and $\hat{i} \cdot \hat{j} = 0$.

Remember the rules on significant figures; in this case, since the vectors as provided could be considered exact values, we follow the precision of the elementary charge.

The charge on an electron is 1.602 × 10^{-19} C, and a joule (kg m^2 s^{-2}) = N m, because the base SI units of the newton are kg m s^{-2}.

? Exercise 6.2 The Vector Dot Product

The six vectors from Exercise 6.1 are listed again:

$$\mathbf{a} = 6\hat{i} + 7\hat{j}$$
$$\mathbf{b} = 7\hat{i} - 5\hat{j}$$
$$\mathbf{c} = -3\hat{j} + 7\hat{k}$$
$$\mathbf{d} = 3\hat{i} - 8\hat{j} - 6\hat{k}$$
$$\mathbf{f} = 9\hat{i} + 7\hat{j} - 2\hat{k}$$
$$\mathbf{g} = -8\hat{i} - 3\hat{j} + 5\hat{k}$$

Evaluate the following dot products and determine the angle between the vectors:

1. $\mathbf{a} \cdot \mathbf{b}$ 4. $\mathbf{g} \cdot \mathbf{d}$
2. $\mathbf{b} \cdot \mathbf{c}$ 5. $\mathbf{c} \cdot \mathbf{f}$
3. $\mathbf{d} \cdot \mathbf{f}$ 6. $\mathbf{f} \cdot \mathbf{b}$

6.4.2   The 'Cross' (Vector) Product

The vector cross product, in contrast to the dot product, is known as the *vector product*, because the result is a vector which is orthogonal to the two originating vectors. The definition of the cross product is shown in Key Concept 6.5; it is found from the magnitudes of the vectors involved and the sine of the angle between them. In contrast to the dot product, however, we now have a unit vector term, $\hat{n}$, indicating a direction orthogonal (perpendicular) to the two component vectors. Because of the sine function, we have a situation where vectors parallel to each other now cancel each other out ($\sin(0) = 0$), whereas vectors perpendicular to each other now multiply together ($\sin(\pi/2) = 1$).

The result of the dot product will always give a scalar value; the result of the cross product will always be a vector. For this reason they are often termed the "scalar product" for the dot product and the "vector product" for the cross product.

---

❶ **Key Concept 6.5 The Vector Cross Product**

The definition of the cross product is:

$$\mathbf{m} \times \mathbf{m} = \hat{\mathbf{n}}|\mathbf{m}||\mathbf{p}|\sin\theta$$

where θ is the angle between the vectors and $\hat{\mathbf{n}}$ is a unit vector *perpendicular* to both originating vectors. The application of the sine function now means that parallel vectors cancel out while perpendicular vectors are directly multiplied together:

$$3\hat{\mathbf{i}} \times 4\hat{\mathbf{i}} = 12\left(\hat{\mathbf{i}} \times \hat{\mathbf{i}}\right) = 12 \cdot (0) = 0$$

$$3\hat{\mathbf{i}} \times 5\hat{\mathbf{j}} = 15\left(\hat{\mathbf{i}} \times \hat{\mathbf{j}}\right) = 15 \cdot (1) = 15\hat{\mathbf{k}}$$

Note now that since we only used $\hat{\mathbf{i}}$ and $\hat{\mathbf{j}}$ unit vectors, the vector which is perpendicular to both of these is $\hat{\mathbf{k}}$.

If two vectors which are perpendicular to each other give another vector, we now need to pay attention to the sign of the result; after all, if we 'cross' $\hat{\mathbf{i}}$ and $\hat{\mathbf{j}}$, there are two possible results: $\hat{\mathbf{k}}$ and $-\hat{\mathbf{k}}$ *By convention*, the direction of the resulting vector is defined as in eqn (6.5). This can be visualised in Figure 6.4, whereby if we 'cross' the vectors in a clockwise direction (with the arrows) we get a positive unit vector result, whereas if we 'cross' the vectors in an anticlockwise direction (against the arrows) we get a negative unit vector result:

$$\hat{\mathbf{i}} \times \hat{\mathbf{j}} = \hat{\mathbf{k}} \quad \hat{\mathbf{j}} \times \hat{\mathbf{i}} = -\hat{\mathbf{k}}$$
$$\hat{\mathbf{j}} \times \hat{\mathbf{k}} = \hat{\mathbf{i}} \quad \hat{\mathbf{i}} \times \hat{\mathbf{k}} = -\hat{\mathbf{j}}$$
$$\hat{\mathbf{k}} \times \hat{\mathbf{i}} = \hat{\mathbf{j}} \quad \hat{\mathbf{k}} \times \hat{\mathbf{j}} = -\hat{\mathbf{i}} \tag{6.5}$$

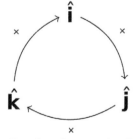

Figure 6.4 Determining the sign of crossed unit vectors: if we follow the arrows ($e.g.\ \hat{\mathbf{j}} \times \hat{\mathbf{k}}$), we have a positive result ($\hat{\mathbf{i}}$), while going against the arrows ($\hat{\mathbf{k}} \times \hat{\mathbf{j}}$) gives a negative result ($-\hat{\mathbf{i}}$).

This *non-commutativity* is an important consideration in vector multiplication (*i.e.* $\mathbf{a} \times \mathbf{b} \neq \mathbf{b} \times \mathbf{a}$).

We can also apply the cross product to find the angle between two vectors in a similar manner as for the dot product, but this time using the definition of the cross product shown in Key Concept 6.5.

> Remember that the sine and cosine functions are cyclic (see Figure 5.6); a number of angles can result in any one value for $\sin \theta$; e.g. a value 0.6 can come from both $\sin(37°)$ and $\sin(143°)$. It can be helpful to make a quick sketch of your vectors to determine which angle is actually needed.

Example 6.6 Determining a Cross Product in Three Dimensions

Q: Find the cross product of the vectors $\mathbf{m} \times \mathbf{p}$. The vectors are defined as

$$\mathbf{m} = -2\hat{\mathbf{j}} + 8\hat{\mathbf{k}}$$

$$\mathbf{p} = 3\hat{\mathbf{i}} - 4\hat{\mathbf{j}}$$

A: While this looks like a two-dimensional problem, we have all of $\hat{\mathbf{i}}, \hat{\mathbf{j}}$ and $\hat{\mathbf{k}}$ vectors present. Let's write the vectors out as a cross product and expand the brackets:

$$\mathbf{m} \times \mathbf{p} = \left(-2\hat{\mathbf{j}} + 8\hat{\mathbf{k}}\right) \times \left(3\hat{\mathbf{i}} - 4\hat{\mathbf{j}}\right)$$

$$= -6\left(\hat{\mathbf{j}} \times \hat{\mathbf{i}}\right) - 32\left(\hat{\mathbf{k}} \times \hat{\mathbf{j}}\right) + 8\left(\hat{\mathbf{j}} \times \hat{\mathbf{j}}\right) + 24\left(\hat{\mathbf{k}} \times \hat{\mathbf{i}}\right)$$

We remember that parallel vectors 'cross' to zero $\left(\hat{\mathbf{j}} \times \hat{\mathbf{j}} = 0\right)$, and for the remaining we consult Figure 6.4. We need to pay attention to whether the 'crossing' goes 'with the arrow' (positive unit vector result) or 'against the arrow' (negative unit vector result):

$$-6\left(\hat{\mathbf{j}} \times \hat{\mathbf{i}}\right) = -6\left(-\hat{\mathbf{k}}\right)$$

$$-32\left(\hat{\mathbf{k}} \times \hat{\mathbf{j}}\right) = -32\left(\hat{\mathbf{i}}\right)$$

$$+8\left(\hat{\mathbf{j}} \times \hat{\mathbf{j}}\right) = +8\left(0\right)$$

$$+24\left(\hat{\mathbf{k}} \times \hat{\mathbf{i}}\right) = +24\left(+\hat{\mathbf{j}}\right)$$

Our result for the cross product $\mathbf{m} \times \mathbf{p}$ is therefore:

$$\mathbf{m} \times \mathbf{p} = +32\hat{\mathbf{i}} + 24\hat{\mathbf{j}} + 6\hat{\mathbf{k}}$$

This vector will be perpendicular to both of the originating vectors $\hat{\mathbf{m}}$ and $\hat{\mathbf{p}}$. We can check this by evaluating the dot product of our result and either of the originating vectors; the answer should come to zero (perpendicular vectors will 'dot' to zero).

❓Exercise 6.3 The Vector Cross Product

Once again we, examine the six vectors from Exercise 6.1:

$\mathbf{a} = 6\hat{\mathbf{i}} + 7\hat{\mathbf{j}}$ $\mathbf{d} = 3\hat{\mathbf{i}} - 8\hat{\mathbf{j}} - 6\hat{\mathbf{k}}$

$\mathbf{b} = 7\hat{\mathbf{i}} - 5\hat{\mathbf{j}}$ $\mathbf{f} = 9\hat{\mathbf{i}} + 7\hat{\mathbf{j}} - 2\hat{\mathbf{k}}$

$\mathbf{c} = -3\hat{\mathbf{j}} + 7\hat{\mathbf{k}}$ $\mathbf{g} = -8\hat{\mathbf{i}} - 3\hat{\mathbf{j}} + 5\hat{\mathbf{k}}$

Find the cross product for each of the following vectors:

1. $\mathbf{a} \times \mathbf{b}$ 4. $\mathbf{g} \times \mathbf{d}$

2. $\mathbf{b} \times \mathbf{c}$ 5. $\mathbf{c} \times \mathbf{f}$

3. $\mathbf{d} \times \mathbf{f}$ 6. $\mathbf{f} \times \mathbf{b}$

You may wish to verify your result for each by finding the 'dot' product of the result with one of the originating vectors; remember that the dot product for perpendicular vectors is equal to zero.

6.4.3   In Context: Dot and Cross Products in Chemical Applications

We showed how the dot product can be used to determine work done (Example 6.5); however, it is useful to show how these products are justified in the context of chemistry. Each of the vector products involve a sine or cosine function, so in our examples we will be using the principles covered in Chapter 5.

6.4.4   Motion in an Electric Field: The Dot Product

Let's revisit the example we first showed in Figure 6.3a; a sodium ion migrating through an electric field. The ion is not migrating fully in the direction of the electric field; however, if we redraw Figure 6.3a, we can use the angle α to determine how far the ion actually moves in the direction of the electric field **E** (Figure 6.5)

The distance the ion moves in the direction of the field is found from $|\mathbf{r}|\cos(\alpha)$ (Figure 6.5b); in Example 6.5, we said that if we multiply this by the electrical force ($\mathbf{F}_{elec} = q\mathbf{E}$) we can calculate the electrical work done. This gives us the calculation for the work done

There is no work done moving perpendicular to the field; as the electrical potential will be unchanged along this line, there is no change in the energy of the system.

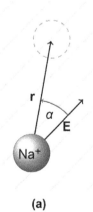

(a)

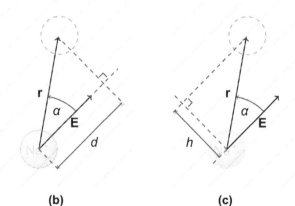

(b)   (c)

Figure 6.5   Considering migration of Na⁺ through an electric field **E** as vectors (a). The distance d the ion moves in the direction of the field **E** (b) is found from $|\mathbf{r}|\cos(\alpha)$, while the distance h moved perpendicular to the field (c) is found from $|\mathbf{r}|\sin(\alpha)$.

(eqn (6.6)), which is the same as the vector dot product for these two vectors (Key Concept 6.4).

$$W_e = |\mathbf{F}_{elec}||\mathbf{r}|\cos(\alpha) = q|\mathbf{E}||\mathbf{r}|\cos(\alpha)$$
$$\equiv \mathbf{F} \cdot \mathbf{r} = q\mathbf{E} \cdot \mathbf{r} \qquad (6.6)$$

This shows that, if we know the vectors involved, we can determine the dot product without needing to know the angle between the two vectors.

6.4.5   *Molecular Torque in an Electric Field: The Cross Product*

Now we shall turn our attention to the problem first shown in Figure 6.3b, but we will consider it in the context of an HCl molecule. Figure 6.6 shows an HCl molecule in the same uniform electric field as before, but now we will consider the torque around the centre of rotation, O.

In chemistry, a molecule with a permanent dipole (which can be represented as a vector) will experience a torque when placed in an electric field. The positive end of the dipole is pulled in the direction of the field while the negative end of the dipole is pulled in the opposite direction. As already mentioned, the torque of a force around a point of rotation is equal to the magnitude of the

> The quantity $\mathbf{r}\cos(\alpha)$ is often termed 'the component of $\mathbf{r}$ in the direction of the electric field', while the quantity $\mathbf{r}\sin(\alpha)$ is 'the component of $\mathbf{r}$ perpendicular to the field'.

> As the work done is a scalar value, we simply use the moduli of the vectors involved.

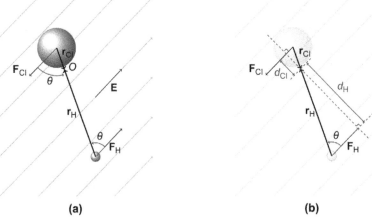

(a)   **(b)**

Figure 6.6   An HCl molecule in a uniform electric field **E**. The electric field induces forces on each atom ($\mathbf{F}_H$ and $\mathbf{F}_{Cl}$), which exert a torque around the centre of rotation, O. The torque of each is related to the perpendicular distance from O to the line of action of each force (d_H and d_{Cl}), determined by $d_x = |\mathbf{r}_x|\sin\theta$.

force multiplied by the perpendicular distance of the force's line of action from the point of rotation.

In the context of Figure 6.6, let's consider the hydrogen atom. It is acted on by force $\mathbf{F}_H$ at angle θ to its position vector $\mathbf{r}_H$ relative to the centre of rotation O. The perpendicular distance of this force's line of action can be calculated as $d_H = |\mathbf{r}_H| \sin\theta$. We can use the same determination for the chlorine atom, and can quantify the torque of each (M_H and M_{Cl}) about the centre as:

$$M_x = |\mathbf{F}_x||\mathbf{r}_x|\sin\theta \tag{6.7}$$

This is starting to look very like the cross product, but we're not quite there yet. We can calculate the torque from each atom in the system using eqn (6.7) and add all the torques; however, there is one last thing to consider—when do torques compete, and when do they combine? This is why we need to consider the *vector* nature of the cross product.

We have already said that the vector cross product is *non-commutative* (Section 6.4.2), and we showed that $\mathbf{a} \times \mathbf{b} = -\mathbf{b} \times \mathbf{a}$ in the context of unit vectors. This is illustrated in Figure 6.7, and gives a way to consider the direction of the resulting vector using a 'right-hand grip'; if we position our right hand so that our fingers follow the direction from the first vector to the second, our thumb will point in the direction of the resulting vector.

Let's now apply this principle to the HCl model in Figure 6.6. For the torque on each atom, we are calculating $\mathbf{F}_x \times \mathbf{r}_x$; if we align our

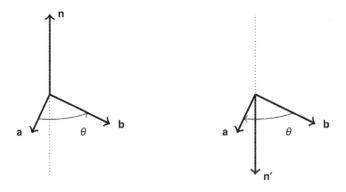

(a) **a × b = n**   (b) **b × a = n′ = −n**

Figure 6.7   Here we illustrate the difference between $\mathbf{a} \times \mathbf{b}$ and $\mathbf{b} \times \mathbf{a}$; if we imagine gripping with right-hand 'thumbs up' position, and allowing our fingers to follow the direction from **a** to **b** (Figure 6.7a), our thumb will point in the direction of the resulting vector **n**. Doing the reverse process (from **b** to **a**, Figure 6.7b), our thumb points in the opposite direction.

right hand on the chlorine atom, and point our fingers in the direction *from* $\mathbf{F}_{Cl}$ *to* $\mathbf{r}_{Cl}$, our thumb will point *upwards* out of the plane of the paper. Doing this again for the hydrogen atom, we should see that our thumb points upwards again; this means that these torques will *add together*, and this should be apparent as both atoms will move anticlockwise around the centre of rotation, O.

We can now finish our expression for the torque around the centre of rotation started in eqn (6.7) to account for its vector nature (eqn (6.9)), incorporating a change to vector notation, $\mathbf{M}_x$:

$$\mathbf{M}_x = \hat{\mathbf{n}}|\mathbf{F}_x||\mathbf{r}_x|\sin\theta$$
$$= \mathbf{F}_x \times \mathbf{r}_x = q_x \mathbf{E}_x \times \mathbf{r}_x \qquad (6.8)$$

Rather than looking at absolute charges in a molecule, we tend to consider the overall *dipole moment*, μ; this is a product of the charge difference across the molecule and the distance over which it acts, with units of C m. As this is a 'charge times distance', we can formulate a new expression for our torque in terms of the dipole moment and electric field (eqn (6.9)):

$$\mathbf{M}_x = \mathbf{F}_x \times \mathbf{r}_x = q_x \mathbf{E}_x \times \mathbf{r}_x$$
$$= \mathbf{E}_x \times q_x \mathbf{r}_x$$
$$= \mathbf{E}_x \times \mu_x \qquad (6.9)$$

Once again, we have demonstrated that using vectors allows us to determine molecular properties without needing to calculate angles in our system.

While this process is straightforward for diatomic molecules, in more complex molecular systems we would need to use computational techniques to determine these torques, and thus the vector approach becomes vital. This model can then be useful in understanding molecular rotations in spectroscopy.

6.5 Vectors in Crystallography

X-ray crystallography is used to study the real structures of atoms and molecules in crystals and is widely used in protein studies, as well as for reporting structures of small molecules. Key to crystallography is the concept of the *unit cell*; the smallest possible repeating structural unit within the crystal. This cell contains all the information on molecular positions within the crystal; being able to handle the mathematics of this allows for computational modelling and predicting of macroscopic properties. This is particularly important in research into photovoltaics for solar power conversion, but is also widely used elsewhere. The principles of

Modelling molecular rotations in this way can be helpful for understanding the nature of rotational spectroscopy; however, it is a gross simplification which ignores the quantum mechanical nature of the transitions.

algebra can then be used to calculate the volume of the unit cell, molecular separations and directions of dipoles within the crystal.

6.5.1 *Vector Products In Context*

Consider a pair of vectors, **m** and **p**, and suppose that these are two position vectors describing a unit cell in two dimensions. We can sketch them, as shown in Figure 6.8a, and we see how these map out the parallelogram base of the crystal cell.

Once again, we have an angle between the vectors, but let's now re-examine the cross product of the vectors. When we consider the sine of angle α and combine it with the magnitude of one of the vectors, we obtain a perpendicular 'height'. Figure 6.8b shows that the height of this parallelogram is determined as $h = |\mathbf{m}|\sin(\alpha)$.

This gives us a final perspective on our vector cross product; two vectors map out a parallelogram, and the cross product of these vectors will give the area of the parallelogram. Interestingly, this means that the area of the parallelogram is also a vector quantity and, if working in three dimensions, this new vector may be used again with the third vector of the unit cell to determine its volume. These vector properties are summarised in Key Concept 6.6.

---

❶ Key Concept 6.6 Geometry of Vector Products

In three dimensions, a set of vectors may be described as:

$$\mathbf{m} = a\hat{\mathbf{i}} + b\hat{\mathbf{j}} + c\hat{\mathbf{k}}$$

$$\mathbf{p} = d\hat{\mathbf{i}} + e\hat{\mathbf{j}} + f\hat{\mathbf{k}}$$

$$\mathbf{q} = g\hat{\mathbf{i}} + h\hat{\mathbf{j}} + n\hat{\mathbf{k}}$$

- The angle between any pair of vectors can be found using the dot (scalar) product of this pair.
- Any **pair** of vectors will enclose a parallelogram; the area of this parallelogram can be found from the cross (vector) product of this pair (*e.g.* $\mathbf{A} = \mathbf{m} \times \mathbf{p}$).
- The vector of this area will be perpendicular to both original vectors.
- The dot (scalar) product of the area with the remaining vector will give the **volume**, V, of the enclosed parallelepiped.
- It does not matter what order the vectors are combined, as the same result will be obtained:

$$V = (\mathbf{m} \times \mathbf{p}) \cdot \mathbf{q}$$

$$= (\mathbf{p} \times \mathbf{m}) \cdot \mathbf{q}$$

$$= (\mathbf{q} \times \mathbf{p}) \cdot \mathbf{m}$$

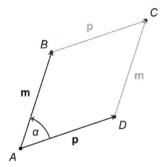
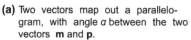
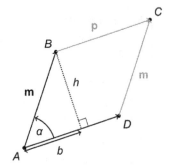

(a) Two vectors map out a parallelogram, with angle α between the two vectors **m** and **p**.

(b) A right-angled triangle can be drawn within the parallelogram.

Figure 6.8   Visualisation of the parallelogram marked out by the vectors **m** and **p** (a). This allows us to gain perspective on the angle α. The 'base' of the triangle, b, is defined by $b = |\mathbf{m}|\cos(\alpha)$, while the height of the parallelogram, h, is defined by $h = |\mathbf{m}|\sin(\alpha)$ (b).

6.6 Summary

In this chapter, we introduced the concept of a vector as a *directed number*, carrying information on *magnitude* as well as *direction*. Directions can be represented in the context of a Cartesian system, using the unit vectors $\hat{\mathbf{i}}, \hat{\mathbf{j}}$ and $\hat{\mathbf{k}}$, which then facilitates the process of vector arithmetic.

In a molecular context, we can determine the overall dipole moment if we know the vectors of the dipole moments for each bond. We can also use vectors to determine physical properties, such as centre of mass, which can help to predict spectroscopic properties. While it is cumbersome to perform such manipulations on large molecules by hand, a computer can do these calculations with ease. Knowing what is going on allows us to make sense of the output of our computer programmes.

As with earlier chapters, we will refer back to this chapter as we progress our discussions of maths in chemistry.

Summary: Chapter 6

- Many quantities in chemistry can be considered as vectors; positions, velocities, dipole moments, torques, angular momentum, forces, fields, *etc.*
- Vector algebra allows for simplification of calculations, negating the need to calculate angles.
- Where parallel dimensions are needed, use the vector dot product.
- Where perpendicular dimensions are needed, use the vector cross product.
- Whether determining a cross or a dot product, both quantities must be vectors. Remember: you cannot cross a scalar with a vector!

Key Equations

For vectors $\mathbf{m} = a\hat{\mathbf{i}} + b\hat{\mathbf{j}} + c\hat{\mathbf{k}}$ and $\mathbf{p} = d\hat{\mathbf{i}} + e\hat{\mathbf{j}} + f\hat{\mathbf{k}}$ with angle θ between them:

- $\mathbf{m} + \mathbf{p} = (a + d)\hat{\mathbf{i}} + (b + e)\hat{\mathbf{j}} + (c + f)\hat{\mathbf{k}}$
- $\mathbf{m} \cdot \mathbf{p} = (ad) + (be) + (cf) = |\mathbf{m}||\mathbf{p}|\cos(\theta)$
- $\mathbf{m} \cdot \mathbf{p} = 0$, if vectors are perpendicular
- $\mathbf{m} \times \mathbf{p} = (bf - ce)\hat{\mathbf{i}} + (cd - af)\hat{\mathbf{j}} + (ae - bd)\hat{\mathbf{k}} = \hat{\mathbf{n}}|\mathbf{m}||\mathbf{p}|\sin(\theta)$
- $\mathbf{m} \times \mathbf{p} = 0$, if vectors are parallel
- $\mathbf{m} \times \mathbf{p} = -\mathbf{p} \times \mathbf{m}$

❓Exercise 6.4 Vectors in a Chemical Context

1. Methanol (CH_3OH) may be modelled with atoms at the following position vectors:

$$\mathbf{r}_C = (0\hat{\mathbf{i}} + 0\hat{\mathbf{j}} + 0\hat{\mathbf{k}}) \text{ Å}$$

$$\mathbf{r}_{H_a} = (-1.03\hat{\mathbf{i}} + 0.37\hat{\mathbf{j}} + 0\hat{\mathbf{k}}) \text{ Å}$$

$$\mathbf{r}_{H_b} = (0.52\hat{\mathbf{i}} + 0.37\hat{\mathbf{j}} + 0.89\hat{\mathbf{k}}) \text{ Å}$$

$$\mathbf{r}_{H_c} = (0.52\hat{\mathbf{i}} + 0.37\hat{\mathbf{j}} - 0.89\hat{\mathbf{k}}) \text{ Å}$$

$$\mathbf{r}_O = (0\hat{\mathbf{i}} - 1.43\hat{\mathbf{j}} + 0\hat{\mathbf{k}}) \text{ Å}$$

$$\mathbf{r}_{H_d} = (0.90\hat{\mathbf{i}} + 1.74\hat{\mathbf{j}} + 0.60\hat{\mathbf{k}}) \text{ Å}$$

Calculate the position vector of the centre of mass.

For Question 2, you will need to determine the unit vector along the C–O bond and scale this by the magnitude of the dipole moment to obtain the vector $\boldsymbol{\mu}$ before using the cross product with the electric field vector. You do not need to adjust this vector for the position of the centre of mass.

2. The dipole moment of methanol is $\boldsymbol{\mu} = 1.70\,D \equiv 5.68 \times 10^{-30}$ C m. Assuming that this dipole acts through the centre of mass in the same direction as the C–O bond and that the molecule is aligned as in Question 6.4, calculate the torque around the centre of mass when the molecule is placed in an electric field $\mathbf{E} = (10\hat{\mathbf{k}})$ N C^{-1}.

For Question 3, before attempting to convert units, you may wish to determine whether this is necessary! $e = 1.602 \times 10^{-19}$ C.

3. A DNA oligonucleotide with a formal charge of $-10e$ migrates across an electric field from its origin to a position vector of $\mathbf{r} = (-4\hat{\mathbf{i}} + 8\hat{\mathbf{j}} - 3\hat{\mathbf{k}})$ cm. Given that $\mathbf{E} = (0.5\hat{\mathbf{i}} + 0.2\hat{\mathbf{j}} - 0.9\hat{\mathbf{k}})$ V cm^{-1}, determine the work done in this transfer, and whether work is done on the DNA or the DNA does work on the surroundings.

Solutions to Exercises

Solutions: Exercise 6.1

1. $\mathbf{a} + \mathbf{b} = 13\hat{\imath} + 2\hat{\jmath} \ (+0\hat{k})$
2. $\mathbf{b} - \mathbf{c} = 7\hat{\imath} - 2\hat{\jmath} - 7\hat{k}$
3. $\mathbf{d} - \mathbf{f} = -6\hat{\imath} - 15\hat{\jmath} - 4\hat{k}$
4. $\mathbf{g} + \mathbf{d} = -5\hat{\imath} - 11\hat{\jmath} - \hat{k}$

5. $\mathbf{c} + \mathbf{f} = 9\hat{\imath} + 4\hat{\jmath} + 5\hat{k}$
6. $\mathbf{f} - \mathbf{b} = 2\hat{\imath} + 12\hat{\jmath} - 2\hat{k}$
7. $3\mathbf{a} + \mathbf{f} = 27\hat{\imath} + 28\hat{\jmath} - 2\hat{k}$
8. $\mathbf{g} - 4\mathbf{b} = -36\hat{\imath} + 17\hat{\jmath} + 5\hat{k}$

Solutions: Exercise 6.2

1. $\mathbf{a}\cdot\mathbf{b} = 7; \theta = 85°$
2. $\mathbf{b}\cdot\mathbf{c} = 15; \theta = 77°$
3. $\mathbf{d}\cdot\mathbf{f} = -17; \theta = 98°$

4. $\mathbf{g}\cdot\mathbf{d} = -30; \theta = 107°$
5. $\mathbf{c}\cdot\mathbf{f} = -35; \theta = 113°$
6. $\mathbf{f}\cdot\mathbf{b} = 28; \theta = 74°$

Solutions: Exercise 6.3

1. $\mathbf{a} \times \mathbf{b} = 0\hat{\imath} + 0\hat{\jmath} - 79\hat{k}$
2. $\mathbf{b} \times \mathbf{c} = -35\hat{\imath} - 49\hat{\jmath} - 21\hat{k}$
3. $\mathbf{d} \times \mathbf{f} = 58\hat{\imath} - 48\hat{\jmath} + 93\hat{k}$

4. $\mathbf{g} \times \mathbf{d} = 58\hat{\imath} - 33\hat{\jmath} + 73\hat{k}$
5. $\mathbf{c} \times \mathbf{f} = -43\hat{\imath} + 63\hat{\jmath} + 27\hat{k}$
6. $\mathbf{f} \times \mathbf{b} = -10\hat{\imath} - 14\hat{\jmath} - 94\hat{k}$

Solutions: Exercise 6.4

1. $\mathbf{r}_c = 0.03\hat{\imath} - 0.74\hat{\jmath} + 0.0\hat{k}$
2. $\mu = -5.68\hat{\jmath} \times 10^{-30}$ C m; $\mathbf{M} = 5.68\hat{\imath} \times 10^{-29}$ N m
3. -3.68×10^{-18} J ; work is done by DNA on the surroundings (it loses energy, shown by negative sign); *i.e.* it migrates spontaneously through the electric field.

Learning Points: What We'll Cover

- Defining the term 'rate of change' in the context of chemical reactions
- Identifying the gradient of a graph as the rate of change
- Rates of change in the context of molecular vibrations
- Introducing the derivative; an algebraic expression to calculate the gradient
- The principles and notation used in calculus for finding the derivative
- Differentiation of polynomials
- Differentiation of special functions; logarithms, exponentials and trigonometric functions
- Using differentiation to find the minimum of a potential energy curve
- Differentiating complex functions: the principles of the chain rule and the product rule
- Differentiating models with multiple variables using the principles of partial differentiation
- Applying these principles to thermodynamics and quantum chemistry

Why This Chapter Is Important

- 'Rates of change' are everywhere in chemistry, from rates of reactions to potential energy curves; we need to be able to handle the mathematics of these changes.
- The rate of change will give us different outcomes depending on the variables involved.
- Different functions change at different rates; any function which can be plotted on a graph has a rate of change.
- The operating principles of calculus are introduced here; these are widely used in the discussion of 'change' in a chemical context.

Calculus 1, Differentiation: Mean Speeds and Equilibrium Separations

Chemistry is often described as a study of change in nature. This highlights the central aspect of a chemical reaction: change. Without change, there would be no chemical reactions for us to study. To understand the nature of these changes and to understand the processes and transformations in our systems, we must understand the *rates of change*; how a concentration varies with time, how a bond strength varies with distance, how enthalpy varies with temperature and so on. We can then use our mathematical models to predict behaviour in chemical systems.

In this chapter, we will introduce the mathematics needed to study the phenomenon of 'change', namely *calculus*. We introduce this in the context of rates of reaction and the potential energy curve of a vibrating diatomic molecule and identify the central mathematical method for determining the rate of change of one variable with respect to another. We then introduce how to apply this to special mathematical functions as well as to chained functions and products of functions. Finally, we show how to apply these principles when we have more than two variables in our system.

7.1 Rates of Change

Rates of change are commonplace in our daily lives, where most of our rates are measured with respect to time, for example a 'rate of change of position with respect to time' we know better as a 'speed'. For a vehicle travelling at 20 km h^{-1}, we expect to see a change in position of 20 km in the time span of 1 h. The arithmetic for such rates of change are likely to be reasonably familiar to you, as shown in Example 7.1.

All rates of change are with respect to another variable, describing how quickly one 'thing' changes with another; it is not always 'with respect to time'.

Note that we observe the rules for significant figures, introduced in Chapter 3.

Speeds are normally reported as km h^{-1} not km min^{-1}.

Units follow the same rules as the calculation—here it was km/h so the unit is km h^{-1}; see Chapter 1 for handling units.

Example 7.1 Speed as 'Rate of Change of Position'

Q: A gas delivery van is stuck in slow-moving traffic and is currently 3.6 km away. You have been informed that the delivery of your argon cylinder will be made in 45 min; what is the mean speed, v, of the delivery van?

A: This is a fairly trivial everyday problem; many will have made mean speed calculations. The speed is simply a *rate of change of position*, and can be calculated in a straightforward manner:

$$v = \frac{3.6\,km}{45\,min}$$

$$= 0.080\,km\;min^{-1} \tag{7.1}$$

$$= 0.080 \times 60\,km\;h^{-1}$$

$$= 4.8\,km\;h^{-1}$$

We can easily interpret this result as 'every hour, the van changes position by 4.8 km'; in this context, this is a mean speed.

The arithmetic for mean speed calculations is straightforward and you will almost certainly be familiar with this particular 'rate of change'. The principles we have introduced are readily applied to rates of change in other contexts.

7.1.1 Rates of Reaction

In chemistry, the first 'rate of change' usually encountered is the 'rate of reaction'. This is most simply defined as how the concentration of a reactant varies with time. Typical undergraduate experiments to determine a reaction rate involve sampling the reaction mixture at given time intervals and recording the concentration at each stage. For a first-order reaction, this process generates data similar to those shown in Figure 7.1.[†]

Such analyses allow us to predict what the concentration of our reactant might be at a given time after the start of the reaction, but they do not directly tell us the rate of the reaction. To determine the rate of reaction at any point from Figure 7.1a, we need to find the rate of change of the reactant with respect to time; this is the *gradient of the graph*. This is done by drawing a *tangent* to the curve, as shown in Figure 7.1b.

[†]Although the rate of reaction is formally a rate of change, it is usually used to formulate an *integrated* form of the rate equation, rather than to determine a value for the rate of reaction at any given time. This is covered in Chapter 8.

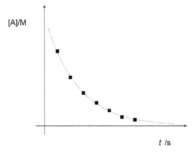

(a) Example concentration data from a kinetics experiment.

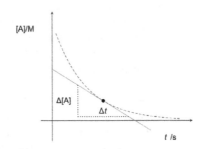

(b) Determining the rate of reaction by calculating a gradient.

Figure 7.1　Example concentration data from a kinetics experiment. Samples are taken at regular intervals and the concentration of the reactant A is measured. The data fit to an exponential decay, indicating a first-order reaction. Having established the fit, a tangent to the curve is drawn at time t and the gradient calculated from the horizontal (Δt) and vertical ($\Delta[A]$) displacements of the tangent at this time. As the gradient changes with time, so the rate of reaction changes as the reaction progresses.

Once we have the tangent to the curve, we measure the change in the vertical dimension ($\Delta[A]$, 'Delta [A]' in this example) and divide it by the change in the horizontal dimension (Δt, 'Delta t'). This will give us the reaction rate:

$$\text{Rate of reaction} = \frac{\text{Change in concentration}}{\text{Elapsed time}} = \frac{\Delta[A]}{\Delta t} \qquad (7.2)$$

This equation gives us the rate of reaction *at the point where the tangent is drawn*, telling us the *rate of change* of one variable (concentration, [A]) *with respect to* another variable (time, t). You will already be familiar with the general form of the equation for a straight line, such as the tangent shown in Figure 7.1b ($y = mx + c$; see Chapter 4), where the gradient m is found from the change in the vertical coordinate, Δy, and the corresponding change in the horizontal coordinate, Δx. This yields the following expression for the gradient, m, of a straight line:

$$m = \frac{\Delta y}{\Delta x} \qquad (7.3)$$

This explains the situation very easily for a straight-line graph; however, many processes in chemistry do not have a straight-line relationship; the first-order reaction shown in Figure 7.1b is only one example of such a process. We therefore need to examine ways of precisely determining our gradients when we do not have a linear relationship between our variables. Let's now look at an example of a non-linear relationship: the harmonic oscillator model for a vibrating diatomic molecule.

7.1.2   *The Harmonic Oscillator: Force Determinations*

Now we know that information is readily obtained by determining the gradient of a graph, it is useful next to cover how to determine a gradient in a more precise manner. In this example, we will consider the determination of a force acting on atoms in a vibrating molecule. The simplest model used to consider a vibrating diatomic is the 'harmonic oscillator' model, which is based on a quadratic formula. A summary of the model is presented in Example 7.2, together with the form of the quadratic equation describing the behaviour of the model.

Example 7.2 The Harmonic Oscillator

The harmonic oscillator model is a first approximation to account for the potential (the energy) associated with the vibrations of a diatomic molecule (e.g. H_2, Cl_2, HCl). It models a chemical bond as a spring between two hard spheres separated by distance r, with an equilibrium bond length of r_0.

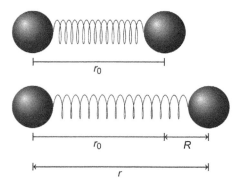

When the bond is stretched by distance R from its equilibrium length, r_0, there is a *restoring force* pulling in the opposite direction to the extension to bring the system back to its equilibrium point. In the absence of any damping, this will result in a vibration and an increase in the energy of the system. Conversely, if the spring is compressed from its equilibrium length, there is another restoring force acting so as to restore the equilibrium separation r_0.

The potential, V, of the harmonic oscillator varies as a function of the extension, R, according to:

$$V_{harmonic} = \frac{kR^2}{2} = \frac{k(r - r_0)^2}{2} \qquad (7.4)$$

where k is the force constant of the spring and R the deviation in length (by either compression or extension) from the equilibrium length. By expanding the brackets in eqn (7.4), we can see the form of the equation in a more familiar setting:

$$V_{harmonic} = \frac{k}{2}\left(r^2 - 2r_0 r + r_0^2\right) \equiv ar^2 - br + c \qquad (7.5)$$

You should recognise this as a simple quadratic equation which, when plotted, will give a parabola, as shown in Figure 7.2 (with parameters $a = k/2\; b = kr_0$ and $c = kr_0^2/2$). This illustrates how much energy is stored in the bond as it is stretched or compressed; the greater the amplitude (greater compression or extension), the more energy is stored in the bond.

Plotting a graph of a quadratic equation results in a parabola (Figure 7.2), and, in the context of the vibrating diatomic molecule, this is a potential energy curve varying as a function of distance. In this case, the rate of change of potential V with respect to distance r will give the *force* acting on the body in question. As we examine the graph, we see that at the equilibrium separation, r_0, the tangent

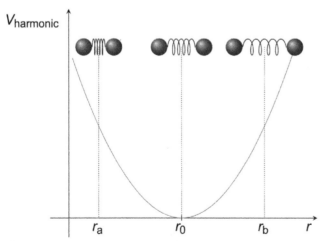

Figure 7.2   The potential of the harmonic oscillator, showing a minimum potential at the equilibrium separation, $r = r_0$. As the bond is compressed ($r_a < r_0$) or stretched ($r_b > r_0$), there is a commensurate increase in the energy of the vibration.

to the curve will have a zero gradient, corresponding to zero force acting on the atoms; this makes sense, as this is an equilibrium! However, should we wish to determine the force at any other point, we would need to draw a tangent to the curve and determine its gradient, as was done for the example in Figure 7.1b.

We can come up with various mathematical methods to approximate a gradient without needing to sketch a tangent; one such method involves drawing tie lines from the point of interest to other points on the graph and systematically reducing the separation between the tie points (Figure 7.3)

> 💬 Tie lines are straight lines connecting two points from which a gradient may be calculated.

We can see from Figure 7.3 that as the tie points get closer to our point of interest at r_1, *i.e.* as $\Delta r \to 0$, the gradient of the cord (line) joining the two points approaches that of the tangent at r_1; therefore as $\Delta r \to 0$, we get closer and closer to determining the force between the two atoms at that distance. Logically, therefore, we will obtain the *exact value for the gradient* when $\Delta r = 0$; *i.e.* when there is zero separation between the two tie points. While this seems like an absurd result (after all, if $\Delta r = 0$, then $\Delta V = 0$, and dividing zero by zero is mathematically problematic), we can explore this in more detail algebraically, described for a general case in Example 7.3.

This process leads us to finding the 'derivative of a function'; a mathematical method which will give the gradient of that function when the gradient triangle used is infinitely small. Derivatives may be calculated mathematically—the rules for doing this are the scope of the rest of this chapter.

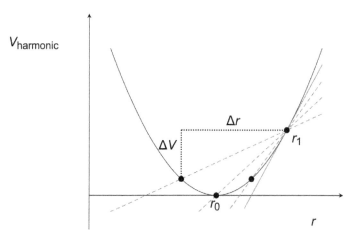

Figure 7.3   The point of interest is connected to successive tie points around the curve; as Δr shrinks, the gradient of the cord joining the tie points approaches that of the tangent at r_1.

Example 7.3 Determining a Gradient

We showed that the harmonic oscillator model may be described as a quadratic relationship (eqn (7.5)). For the purposes of this exercise, we will simplify this to a general quadratic equation. You will find it useful to work through the steps one-by-one; equip yourself with a pen and paper and ensure you are able to follow the steps below.

Consider the relationship in eqn (7.5), which follows the form

$$y = ax^2 - bx + c \qquad (7.6)$$

1. It often helps to begin by sketching the form of this function. You may find it useful to do this (substitute values for a, b and c to do this if it helps).

   What is the gradient between two points: x and $(x + \delta x)$? Remember that δx is the change in x, so we can now use this to find δy; the *change in y*.

2. Find the new value $(y + \delta y)$ by substituting $(x + \delta x)$ into eqn (7.6):

$$(y + \delta y) = a(x + \delta x)^2 - b(x + \delta x) + c \qquad (7.7)$$

3. Expand the brackets in eqn (7.7). (See Toolkit A.2 if necessary.)
4. Subtract eqn (7.6) from eqn (7.7). (See Toolkit A.3 if necessary.)
5. Ensure that δy is the subject of the equation (rearranging if needed; see Chapter 1).
6. Find the gradient $\delta y / \delta x$ by dividing both sides by δx.

   You should now have an equation in the form

$$\frac{\delta y}{\delta x} = 2ax - b + a\delta x \qquad (7.8)$$

7. If we now use the notation dx to indicate that δx has become so small that it is effectively zero, which mathematically we say 'in the limit as $\delta x \to 0$' (see Key Concept 7.1), this equation becomes

$$\frac{dy}{dx} = 2ax - b \qquad (7.9)$$

8. Bringing this back to the harmonic oscillator in eqn (7.5), we can replace the constants a and b from eqn (7.9) to determine the attractive force, F, between the two particles:

$$V_{\text{harmonic}} = \frac{k}{2}\left(r^2 - 2r_0r + r_0^2\right) \equiv ax^2 - bx + c$$

Therefore:

$$\frac{dV(r)}{dR} = -F = -\left(kr - kr_0\right) = -kR \qquad (7.10)$$

Note that, by convention, a restoring force is negative with extension (the force is directed in the opposite direction to the extension), requiring the negative component of eqn (7.10).

The principles covered in Example 7.3 are central to the mathematical technique called *calculus*. Calculus is the name given to the branch of mathematics which is used to describe and model outcomes based on continual change, making it ideally suited to chemistry.

An infinitesimal is a number which is infinitely small—but greater than zero—such that there is no number between it and zero. This is a mathematical construct and there is no need for chemists to worry further.

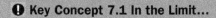

❶ Key Concept 7.1 In the Limit...

'*In the limit*' is a key idea in calculus, and is usually used in the form of

In the limit as $\delta x \rightarrow 0, \delta x = dx$.

It is used to mean 'As the value of δx gets as close to zero as it is possible to get, without being equal to zero'. This is the idea of an *infinitesimal*, and carries the notation dx ('dee x'). This is distinct from upper-case Δx ('Delta x', an observable change) and lower-case δx ('delta x'; a small finite change).

7.1.3 Differentiation—Finding the Gradient of a Function

The process we have gone through is called *differentiation* and allows us to obtain an expression for the gradient at any point on a given curve. The expression dy/dx (often said '*dee y by dee x*') is called the *first derivative of y with respect to x*. This, of course, will depend on our variables, with '*dee a by dee b*' and '*dee s by dee t*' showing rates of change of other variables (not just y and x).

The units of this first derivative are the units of y divided by the units of x.

When we compare eqn (7.6) and (7.9) in Example 7.3, we notice the following trends between the original function and the calculated derivative:

- In the first term, you should notice that the *power has dropped by one* on finding the first derivative; the power of '2' has become a power of '1'.
- In the first term, you should also notice that it has been scaled by the magnitude of the original index; where the first term was originally a power of two ('x^2'), in the derivative, this has become a power of one multiplied by two ($2x^1 \equiv 2x$).
- This same trend is then shown by the next term in the expression; the power has dropped by one and the whole term multiplied by the original power.

These results allow us to formulate a general method for finding a derivative of a given function:

- In general terms, x^n becomes nx^{n-1}; so x^3 when *differentiated* becomes $3x^2$, and so on for other powers. This is true as long as $n \neq 0$, so for negative indices we still follow the rule, and the derivative of x^{-n} with respect to x is $-nx^{-n-1}$, so x^{-2} becomes $-2x^{-3}$.

- Any numerical constants (*i.e.* no component of the variable) can be considered as being a 'zeroth power' of the variable, *i.e.* $5 \equiv 5x^0$; when we differentiate this, the whole value is scaled by zero, and so constants disappear from the derivative. This has consequences later on when we try to reverse the process in Chapter 8.

The process of differentiation, or *finding the derivative*, is vital in chemistry, as it allows us to explore the relationships and dependencies between variables, provided we have a mathematical expression to describe the system.

> When anything is raised to the power 0, the value is 1 ($x^0 = 1$) and so it differentiates as a constant (and goes to 0).

❗ Key Concept 7.2 Differentiation: Simple Functions

The general process of handling a single-term function is described in the following manner.

For the general relationship:

$$y = Ax^b \tag{7.11}$$

1. Multiply the right-hand side by the power of x, and reduce the power of x by one:

$$\frac{dy}{dx} = b \times Ax^{b-1} \tag{7.12}$$

2. Simplify:

$$\frac{dy}{dx} = bAx^{b-1} \tag{7.13}$$

Where there are several functions of x added together, simply differentiate each one separately:

$$y = Ax^b + Cx^g \tag{7.14}$$

$$\frac{dy}{dx} = bAx^{b-1} + gCx^{g-1} \tag{7.15}$$

3. For a negative power of x (see Chapter 1 for the meaning of negative indices), the same rules still apply:

$$y = Ax^{-b} \tag{7.16}$$

$$\frac{dy}{dx} = -bAx^{(-b-1)} \tag{7.17}$$

❓Exercise 7.1 Finding the Gradient of a Function

Using the method laid out in Key Concept 7.2, for each of the following equations derive an expression for the gradient shown.

1. $a = 5b + 3$; $\dfrac{da}{db}$

2. $x = \dfrac{z^3}{6}$; $\dfrac{dx}{dz}$

3. $s = t^2 + 2t + 1$; $\dfrac{ds}{dt}$

4. $y = 3x^2 + 2x - 1$; $\dfrac{dy}{dx}$

5. $f = 1.25g^4 + 0.5g^2$; $\dfrac{df}{dg}$

6. $p = 2m^3 + 6m + 7$; $\dfrac{dp}{dm}$

7. $n = \dfrac{p^4}{2} + \dfrac{p^2}{4}$; $\dfrac{dn}{dp}$

8. $c = 2(b^3 + b + 7)$; $\dfrac{dc}{db}$

Note: We have deliberately used different symbols to represent different variables, but the underlying principles are always the same.

7.1.4 Alternative Notation

Alternative notation systems exist; these are summarised in Table 7.1. Leibniz's notation is most often used in the physical sciences as it is the most versatile, while Lagrange's notation is frequently used to make it absolutely clear which term is the variable. Newton's notation is very rarely used and only included for the sake of completeness. Remember that in *plain English*, the notation is simply stating:

... the rate of change of [y] with respect to [x] is equal to ...

If our variables instead were pressure, p, and volume, V, then we would say:

$\dfrac{dp}{dV}$, *the rate of change of pressure with respect to volume is ...*

Table 7.1 A summary of different notations in calculus. Successive derivatives are found using the approach described in Key Concept 7.2.

	Leibniz's notation	Lagrange's notation	Newton's notation
Function	$y = Ax^b$	$f(x) = Ax^b$	Ax^b
First derivative	$\dfrac{dy}{dx} = bAx^{b-1}$	$f'(x) = bAx^{b-1}$	$\dot{x} = bAx^{b-1}$
Second derivative	$\dfrac{d^2y}{dx^2} = b(b-1)Ax^{b-2}$	$f''(x) = \ldots$	$\ddot{x} = \ldots$
Third derivative	$\dfrac{d^3y}{dx^3} = b(b-1)(b-2)Ax^{b-3}$	$f'''(x) = \ldots$	$\dddot{x} = \ldots$

In chemistry, we must be able to handle a wide range of similar problems; however, the variables will not always be concentration or time. That said, the maths surrounding the problem is exactly the same, regardless of the symbols carried by the variables.

❓Exercise 7.2 More Differentiation Practice

In this exercise, practise your skill of differentiation using the techniques introduced in Key Concept 7.2 to find the derivative specified after each example. You may assume that all other terms in the given equations are constant. We have also implemented some new notation; question 1 is the same as for Exercise 7.1 to give some familiarity with the different notation.

1. $f(b) = 5b + 3$; $f'(b)$

2. $g = 3f^2 - 6$; $\dfrac{dg}{df}$

3. $f(t) = 8t + \dfrac{3t^2}{2}$; $f'(t)$

4. $f(a) = 3a^9 + \dfrac{3}{a^2}$; $f'(a)$

5. $f(b) = 3b^{-3} + 2b^5 - 5b^{-8}$; $f''(b)$

6. $x = 4y^{1/2} - 5$; $\dfrac{dx}{dy}$

7. $f(y) = 9x^{2/3} - 3x^{1/3}$; $f'(y)$

8. $z = 8a^{0.8} - 6a^{-0.4}$; $\dfrac{dz}{da}$

9. $pV = nRT$; $\dfrac{dp}{dT}$

10. $G = H - TS$; $\dfrac{dG}{dT}$

> 📎 For Question 9, you will need to rearrange this equation to make p the subject.

7.2 Differentiating Special Functions

We have seen how to differentiate simple polynomial expressions, but we also need to differentiate other functions in chemistry. Many expressions in reaction kinetics and thermodynamics involve either an exponential function (in the form e^{Ax}) or a logarithmic function (in the form $\ln(Bx)$). When considering wave functions in quantum chemistry, we need to use a wave description of electrons, requiring the use of trigonometric functions in the form of $\sin(kx)$ and $\cos(kx)$. Each of these functions can be plotted as a graph, which means that each will have an associate gradient and hence a first derivative with particular meaning:

- A plot of reactant concentration against time for a first-order reaction typically has a negative exponential relationship between the two; the gradient will tell us the *rate of reaction*.
- A wave function describes a particle's motion through space; the *gradient* of the wave function (the first derivative) allows the momentum to be determined, while the *curvature* of this wave function (its second derivative) can be used to quantify the kinetic energy of the particle.

> 💬 The 'negative' exponential represents a decay in concentration.

7.2.1 Exponential Functions

> ❶ **Key Concept 7.3 Differentiating e^x**
>
> Differentiating the exponential function e^x is, at first glance, extremely simple, as the function is unchanged:
>
> $$y = e^x$$
>
> $$\frac{dy}{dx} = e^x \tag{7.18}$$
>
> For higher powers of e^x, we need to make the following change:
>
> $$y = e^{Ax}$$
>
> $$\frac{dy}{dx} = Ae^{Ax} \tag{7.19}$$
>
> Note that, unlike polynomials (x^3 for example), the index of e^{Ax} is unchanged by the process of differentiation. We can also demonstrate the differentiation of e^{Ax} using the *chain rule*, covered in Key Concept 7.6, and using the substitution $u = Ax$.

> Remember that fractions containing an index can be rearranged, where
> $$e^{-a} = \frac{1}{e^a}.$$

❓Exercise 7.3 Differentiating Exponential Functions

1. $a(b) = e^{2b}$; $a'(b)$

2. $y = 3e^{2x} + 2x - 1$; $\dfrac{dy}{dx}$

3. $r = e^{2s} + 2e^{-2s}$; $\dfrac{dr}{ds}$

4. $p(q) = \dfrac{4}{e^{7q}}$; $p'(q)$

5. $f = \dfrac{e^{-g/4}}{2}$; $\dfrac{df}{dg}$

6. $[A] = [A]_0 \, e^{-kt}$; $\dfrac{d[A]}{dt}$

Note: We have deliberately used different symbols to represent different variables, but the underlying principles are always the same.

7.2.2 Logarithmic Functions

> We can differentiate $\log_{10}(x)$ by converting this to a natural logarithm and then differentiating, as shown here. See the rules of logs, given in Chapter 1, for conversion to the natural logarithm.

> ❶ **Key Concept 7.4 Differentiating $\ln x$**
>
> As you will recall from Chapter 1, the *natural logarithm*, $\ln x$, is the inverse function of the exponential function, e^x. To differentiate this, we must go through a slightly different process to yield our result:
>
> $$y = \ln x$$
>
> $$\frac{dy}{dx} = \frac{1}{x} \tag{7.20}$$
>
> For multiples of x, the following applies:
>
> $$y = \ln(Ax)$$

$$= \ln A + \ln x$$

$$\frac{dy}{dx} = 0 + \frac{1}{x} = \frac{1}{x} \qquad (7.21)$$

For powers of x, the following applies:

$$y = \ln\left(x^B\right)$$

$$= B \ln x$$

$$\ln\left(a^b\right) = b \ln a.$$

$$\frac{dy}{dx} = B \times \frac{1}{x} = \frac{B}{x} \qquad (7.22)$$

These can also be demonstrated using the chain rule, covered in Key Concept 7.6.

? Exercise 7.4 Differentiating Logarithmic Functions

1. $a = 5 \ln b$; $\quad \dfrac{da}{db}$ 4. $p = \ln\left(\dfrac{q^2}{q_0}\right)$; $\quad \dfrac{dp}{dq}$

2. $y(x) = \ln(7x)$; $\quad y'(x)$ 5. $f(g) = 0.5 \ln\left(3g^4\right)$; $\quad f'(g)$

3. $r = \ln\left(s^2\right) + e^{-2s}$; $\quad \dfrac{dr}{ds}$ 6. $y = \ln(2z) - \ln 2 + \ln\left(z^2\right)$; $\quad \dfrac{dy}{dz}$

For help with how to handle logarithms, see Chapter 1.

7.2.3 Trigonometric Functions

The sine and cosine functions are intrinsically related, and this is demonstrated further when finding the first derivative of each. The first derivative of the sine function is shown in eqn (7.23), while the first derivative of the cosine function is shown in eqn (7.24):

$$y = \sin x$$

$$\frac{dy}{dx} = \cos x \qquad (7.23)$$

$$y = \cos x$$

$$\frac{dy}{dx} = -\sin x \qquad (7.24)$$

For multiples of x, the cycle of differentiation proceeds as:

$$y = \sin Ax$$

$$\frac{dy}{dx} = A \cos Ax$$

$$\frac{d^2y}{dx^2} = -A^2 \sin Ax \qquad (7.25)$$

$$\frac{d^3y}{dx^3} = -A^3 \cos Ax$$

The origin of these derivatives comes from the Maclaurin expansion for trigonometric functions but its elaboration is outwith the scope of this text. You may find it interesting to explore the derivation, however a full explanation as to why these derivatives are what they are is given in Toolkit A.12.

> ❗ **Key Concept 7.5 Differentiating sin x and cos x**
>
> Sine and cosine functions differentiate in a cycle, where each of the sine and cosine functions alternate their sign and their nature:
>
>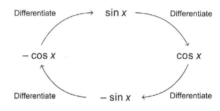

❓Exercise 7.5 More Differentiation

Find the labelled derivative for each of the following functions. Make sure that you pay attention to the sign of the function.

1. $a = 5e^{3b}$; $\dfrac{da}{db}$

2. $y = 5\sin x$; $\dfrac{dy}{dx}$

3. $f(x) = 18e^{2x} + 2\ln(4x)$; $f'(x)$

4. $r(s) = s^{-2} + s^{-0.5}$; $r'(s)$

5. $z = ae^{-ay} + be^{by}$; $\dfrac{dz}{dy}$

6. $g(a) = -3\sin(3a) + 4\cos(2a)$; $g'(a)$

7. $\phi = 8\cos(k\theta)$; $\dfrac{d\phi}{d\theta}$

8. $\Psi_x = Ae^{bx}$; $\dfrac{d\Psi}{dx}$

9. $V(r) = \dfrac{q_1 q_2}{4\pi\varepsilon_0 r}$; $V'(r)$

10. $V(r) = 4\varepsilon\left[\left(\dfrac{r_0}{r}\right)^{12} - \left(\dfrac{r_0}{r}\right)^{6}\right]$; $\dfrac{dV(r)}{dr}$

7.3 Finding the Maximum and the Minimum

We have explored how the mathematical process of *differentiation* allows us to determine an expression to find the gradient of a curve at any point. This can be reversed to find the point at which a gradient is a specific value. This is most commonly used to find the point at which the gradient is zero, where *the rate of change is zero*. In chemistry, we can identify a point of maximum molecular velocity, a point of minimum potential for a molecular system, the maximum rate of reaction, and so on. Notice that all of these phrases denote a

💬 You will often see the terms 'stationary point' and 'turning point' used interchangeably. A *turning point* refers to either a maximum or a minimum; points where the gradient changes its sign. These are also stationary points, but a stationary point can also refer to a point of inflection where the gradient does not change its sign.

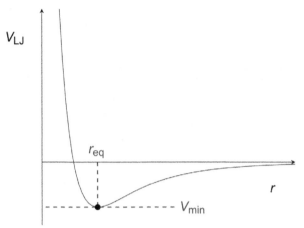

Figure 7.4 Identifying the minimum point of the Lennard-Jones potential (V_{LJ}) for intermolecular forces. The plot shows the variation of the intermolecular potential with intermolecular separation r; the point at which the gradient is zero $\left(\mathrm{d}V_{LJ}/\mathrm{d}r = 0\right)$ identifies the point at which the force is zero.

maximum or a *minimum*—at these *turning points*, the tangent is fully 'horizontal', with a gradient of zero, as shown in Figure 7.4.

It is this ability to determine a *turning point* which makes differentiation so powerful in the physical sciences; in a potential energy curve, a minimum can identify a stable state, while a maximum can show a transition state between two stable states. Remember that the gradient of any graph must have appropriate consideration of units, as discussed in Chapters 1 and 4; for the Lennard-Jones potential shown in Figure 7.4, the gradient will carry the unit of force, the newton (N). It is worth taking the time to ensure that you are happy with this result by taking the SI units for energy (J) and distance (m) and confirming this result. You will need to express the newton and joule in base units to validate this (see Section 1.4).

7.3.1 The Lennard-Jones Potential

The Lennard-Jones potential is expressed in many ways; for the purposes of this text, we have simplified the constants into two values, A and B:

$$V_{LJ} = -\frac{A}{r^6} + \frac{B}{r^{12}} \quad (7.26)$$

The equation describes how the potential, V_{LJ}, varies as a function of intermolecular separation, r; this is shown in Figure 7.4. The overall potential is the result of adding together the repulsive $\left(B/r^{12}\right)$ and attractive $\left(-A/r^6\right)$ potentials. The constants A and B include the equilibrium separation between the molecules (r_{eq}) and the value of the potential at this separation (V_{min}). These are the coordinates corresponding to the minimum point shown in Figure 7.4.

Remember that, by convention, attractive potentials are negative. Potential is defined as zero at infinite separation, with all potential changes of a molecule defined as the 'work done in removing a particle to infinite separation'. If there is attraction, thermodynamic work must be put into the system, increasing its energy; therefore, attractive potentials must be less than zero.

Example 7.4 The Minimum of the Lennard-Jones Potential

To find an expression for the minimum of the Lennard-Jones potential, we need to find its first derivative, and set this to zero. First, recall the form of the Lennard-Jones potential (eqn (7.26)). To simplify the method outlined in Key Concept 7.2, it is easier to write the X/r^n terms as negative indices (Xr^{-n}) instead:

$$V_{LJ} = \frac{A}{r^6} + \frac{B}{r^{12}} \equiv -Ar^{-6} + Br^{-12} \qquad (7.27)$$

Now that we have simplified the layout of the equation, we can apply the method from Key Concept 7.2 to eqn (7.27):

$$\frac{dV_{LJ}}{dr} = (-6) \times -Ar^{(-6-1)} + (-12) \times Br^{(-12-1)}$$

$$= 6Ar^{-7} - 12Br^{-13} \qquad (7.28)$$

To find the point at which the gradient is zero, *i.e.* the value of r when the potential is at a *minimum* (the turning point), we simply set this first derivative in eqn (7.28) to zero when $r = r_{eq}$, and rearrange to obtain an expression for r_{eq}:

$$6Ar_{eq}^{-7} - 12Br_{eq}^{-13} = 0$$

This may be rearranged (see Chapter 1) to:

$$r_{eq} = \left(\frac{2B}{A}\right)^{1/6} \qquad (7.29)$$

This final expression allows us to calculate the equilibrium separation between molecules in terms of A and B for any given molecular system.

There may be more than one solution!

You will need to factorise some of the first derivatives, as shown in the example.

❓Exercise 7.6 Finding the Turning Points of Functions

For each of the following examples, find the first derivative and use this to identify the coordinates at which the gradient is equal to zero:

1. $a = 5b^2 + 6b$

2. $g = 3h^2 - 2h$

3. $p = 3a^3 + 3a^2$

4. $y = 32x^{-2} + 8x$

5. $g = h^3 - 3h + 2$

6. $l = (-1 - 3m^2)(4 - 4m)$

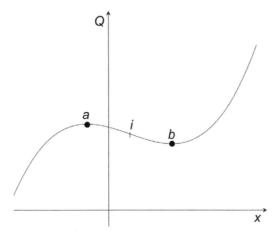

Figure 7.5 The turning points, *a* and *b*, are where the gradient of the curve (the *first derivative*) is equal to zero. Point *i* is the *point of inflection*. At this point the *second derivative* of the curve is equal to zero.

7.3.2   *The Nature of the Turning Point: Maximum or Minimum?*

It is not possible to tell the nature of the turning point from the first derivative alone; the first derivative only allows us to find where the turning point is and cannot tell us whether it is a maximum, a minimum or a point of inflection (Figure 7.5). A potential energy curve can have a number of turning points, corresponding variously to maxima ('unstable' states) and minima ('stable' states); being able to identify the nature of the turning point allows us to qualify whether a state is stable or unstable.

To determine the nature of the turning point, we use the *second derivative*, as this allows us to find the *rate of change of the gradient*. An explanation of this is offered in Example 7.5, and is readily applied to cases of the maximum and minimum.

Example 7.5 Finding a Maximum Using the Second Derivative

Consider the curve plotted in Figure 7.6; we will consider two other points on this curve and examine how the gradient changes.

We can also examine this mathematically; the form of the graph in Figure 7.6 is:

$$f(x) = -ax^2 + bx + c$$

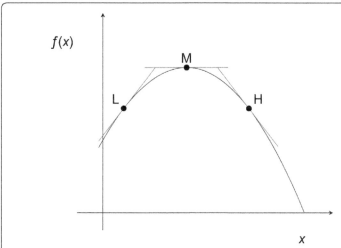

Figure 7.6   Showing how the gradient of a graph moves over a maximum. At a lower value of the variable x, the gradient ($f'(x)$) at point L is greater than zero. At the maximum, M, the gradient is zero ($f'(x) = 0$) and at a higher value of x, at point H, the gradient is negative ($f'(x) < 0$).

The first derivative of this, $f'(x)$ is found using the general method for differentiation:

$$f'(x) = -2ax + b$$

The second derivative is found by differentiating again:

$$f''(x) = -2a$$

We can see that for a graph of this form, the second derivative, or the *rate of change* of *the gradient* is constant and negative. This makes sense when looking at Figure 7.6; the gradient continuously gets smaller ('less steep') as x increases until it reaches zero at the maximum point, M, and then becomes more and more negative as x continues to increase. This is a key result in determining the nature of a stationary point.

A similar principle may be applied for finding minima and points of inflection:

$$f''(x) < 0 \quad \text{Maximum}$$
$$f''(x) > 0 \quad \text{Minimum} \qquad (7.30)$$
$$f''(x) = 0 \quad \text{Point of inflection}$$

If you still have a variable in your secondary derivative (e.g. $f''(x) = 5x$), you will need to put the value of x at the stationary point into the equation to determine whether it is positive or negative. In this example, for a stationary point at $x = 2$, $f''(x) > 0$, which indicates a minimum, while if the stationary point is at $x = -1$, $f''(x) < 0$, which indicates a maximum.

?Exercise 7.7 Finding the Nature of Turning Points

For each of the turning points determined in Exercise 7.6, find the second derivative and use this, together with the coordinates found in Exercise 7.6, to determine whether the turning point is a maximum, a minimum or a point of inflection. These functions may have more than one turning point; you should aim to find the nature of all turning points.

Remember a turning point can be a stationary point, but not all stationary points are turning points.

7.4 More Advanced Differentiation

We have seen the simple application of differentiation for additive functions (polynomials) of the general form $f(x) = ax^n + bx^{n+1} + \cdots$; we simply differentiate each term separately and add the result together. However, many relationships in chemistry are *compound functions*, where different functions of the variable are either multiplied together (*e.g.* $f(x) = x \sin(2x)$) or 'nested' (*e.g.* $f(x) = e^{1/x^2}$). To handle a full range of mathematical models in chemistry, we must be able to fully differentiate such functions using a variety of tools. The two we will cover here are the *chain rule* and the *product rule*.

7.4.1 Chain Rule: The Morse Potential

The Morse potential is another model to describe the interaction between atoms in a diatomic molecule. It is a more 'correct' model as, unlike the harmonic oscillator model, it considers that at extreme compression the nuclear repulsion dominates, while at extreme separation the bond will eventually break. As with the Lennard-Jones potential, the Morse potential is also zero at infinite separation, dropping to a minimum at the equilibrium bond length. The potential is shown in Figure 7.7.

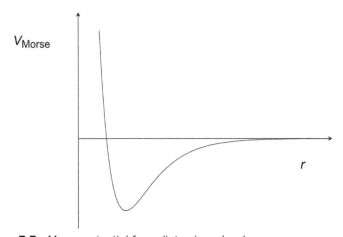

Figure 7.7 Morse potential for a diatomic molecule.

On initial inspection of the plot, the Morse potential appears to be the same as the Lennard-Jones potential (Figure 7.4); however, the form of the equation is very different

$$V_{\text{Morse}} = D \times \left(1 - e^{\left[-a\left(r - r_{\text{eq}}\right)\right]}\right)^2 \tag{7.31}$$

The Morse potential (V_{Morse}) varies with interatomic separation (r), incorporating the system characteristics of 'well depth' D (the value of the potential at the equilibrium separation, r_{eq}) and the 'well width' a (incorporating the masses of the two atoms and the force constant of the bond).

To find the first derivative, we need to use a different approach. As the variable r is part of another function within an exponential, we must find a way to treat each function separately so that we may recombine them. This tool is called the *chain rule*, and is demonstrated for a general function in Example 7.6.

❗ **Key Concept 7.6 Chain Rule—Differentiating Compound Functions**

A compound function takes the form:

$$y = f\left(g[x]\right)$$

Its first derivative may be found using the substitution $u = g(x)$; thus $y = f(u)$, and applying the principle:

$$\frac{dy}{dx} = \frac{dy}{du} \times \frac{du}{dx} \tag{7.32}$$

The chain rule can be extended by as many substitutions as needed to solve the problem:

$$\frac{dy}{dx} = \frac{dy}{du} \times \frac{du}{dt} \times \frac{dt}{ds} \times \frac{ds}{dx} \tag{7.33}$$

The term 'chain' comes from the 'chain of cancellation' arising from treating the respective rates of change as fractions:

$$\frac{dy}{du} \times \frac{du}{dt} \times \frac{dt}{ds} \times \frac{ds}{dx} = \frac{dy}{dx}$$

It does not matter how many substitutions we make in our original compound function, and we can make whatever substitutions we wish, as long as we can find each derivative in the chain:

$$\frac{d(\text{rhubarb})}{d(\text{custard})} = \frac{d(\text{rhubarb})}{d(\text{sugar})} \times \frac{d(\text{sugar})}{d(\text{elephants})} \times \frac{d(\text{elephants})}{d(\text{custard})}$$

Example 7.6 Application of Chain Rule to a Typical Function

When functions are nested, it is possible to find the derivative by substitution. Consider the following example:

$$y = e^{2x^2}$$

It is not possible to find the first derivative directly. Instead we apply a substitution.

$$\text{Let } u = x^2$$

$$\text{Then } y = e^{2u}$$

We can now differentiate y with respect to u (i.e. find dy/du):

$$\frac{dy}{du} = 2e^{2u} = 2e^{2x^2}$$

We can also differentiate u with respect to x (i.e. find du/dx):

$$\frac{du}{dx} = 2x$$

As the expression for the gradient, dy/dx, is a fraction, we can apply similar logic:

$$\frac{dy}{du} \times \frac{du}{dx} = \frac{dy}{d\!\!\!/u} \times \frac{d\!\!\!/u}{dx} = \frac{dy}{dx} \qquad (7.34)$$

Therefore, for this example, the first derivative is:

$$\frac{dy}{dx} = \frac{dy}{du} \times \frac{du}{dx} = 2e^{2x^2} \times 2x = 4xe^{2x^2}$$

> A number of substitutions can be used for this example; the substitution $u = 2x^2$ ($y = e^u$) would be an equally valid substitution and will yield the same result.

> Remember that you can use any substitution you wish; however, the substitution you choose should be differentiated simply, using the rule introduced in Key Concept 7.2.

? Exercise 7.8 Application of the Chain Rule

Use the chain rule to find the derivatives of the following examples:

1. $f(a) = \sin(a^2)$; $\qquad f'(a)$
2. $y = 4\cos(5x^3)$; $\qquad \dfrac{dy}{dx}$
3. $f(b) = e^{\sin(5b)}$; $\qquad f'(b)$
4. $c = -\sin\left(\dfrac{3}{b} + b^2\right)$; $\qquad \dfrac{dc}{db}$
5. $y = e^{(5/x^2)}$; $\qquad \dfrac{dy}{dx}$
6. $f(v) = e^{(v^3/2)}$; $\qquad f'(v)$

> Having made a substitution, you may find it helpful to expand any brackets.

? Exercise 7.9 The Minimum Value of the Morse Potential

Having practised using the chain rule, identify an appropriate substitution and use the chain rule to find the first derivative of the Morse potential:

$$V_{\text{Morse}} = D \times \left(1 - e^{[-a(r - r_{eq})]}\right)^2$$

> You may wish to make a second substitution instead of expanding brackets.

7.4.2   Product Rule: The Maxwell–Boltzmann Distribution

The Maxwell–Boltzmann distribution of molecular speeds, central to the kinetic theory of molecular interactions, illustrates the effect of temperature and molecular mass on the distribution of kinetic energy in a system. The derivation of the function (shown in eqn (7.35)) is a consequence of considering the kinetic energy of any one molecule in three dimensions and multiplying by Boltzmann's energy distribution, which is why the function has its compound form. The plot of the distribution is shown in Figure 7.8.

Specifically, the equation of the distribution (eqn (7.35)) has two functions multiplied together; the first in the form av^2 and the second in the form e^{-Bv^2}.

$$f(v) = 4\pi \left(\frac{m}{2\pi k_B T} \right)^{3/2} v^2 e^{\left(-\frac{mv^2}{2k_B T} \right)} \tag{7.35}$$

The distribution describes the effect that mass (m) and temperature (T) have on the distribution of molecular speeds, v. To find the first derivative, we use the *product rule* for finding the derivative of functions multiplied together. This is demonstrated for a general function in Example 7.7.

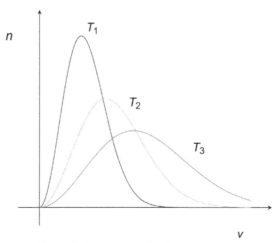

Figure 7.8   The Maxwell–Boltzmann distribution of molecular velocities, showing the broadening of the distribution as temperature increases from T_1 to T_3.

> ❶ **Key Concept 7.7 Product Rule—Differentiating Compound Functions**
>
> For a compound function of the form:
>
> $$y = uv$$
>
> where u and v are different functions of x, the first derivative may be found using the general differential form for products:
>
> $$d(uv) = vdu + udv \qquad (7.36)$$
>
> This is applied as:
>
> $$\frac{dy}{dx} = v\frac{du}{dx} + u\frac{dv}{dx} \qquad (7.37)$$
>
> There are a number of proofs for the product rule; however, the details are outwith the scope of this text.

Example 7.7 Application of Product Rule to a Typical Function

When functions are multiplied together, it is possible to apply the product rule to find the derivative. Consider the following example:

$$y = x^2 e^{3x}$$

Once again, it is not possible to find the derivative directly, so we need to identify the two functions in the composite function and find the derivative of each:

$$u = x^2 \qquad \frac{du}{dx} = 2x$$

$$v = e^{3x} \qquad \frac{dv}{dx} = 3e^{3x}$$

We now apply the product rule to determine the overall derivative for this example:

$$\frac{dy}{dx} = v\frac{du}{dx} + u\frac{dv}{dx}$$

$$\frac{dy}{dx} = \left[e^{3x} \times 2x\right] + \left[x^2 \times 3e^{3x}\right]$$

$$= 2xe^{3x} + 3x^2 e^{3x}$$

$$= xe^{3x}(2 + 3x) \qquad (7.38)$$

? Exercise 7.10 Application of the Product Rule

Use the product rule to find the derivatives of the following examples.

1. $a = b\sin(2b)$; $\dfrac{\mathrm{d}a}{\mathrm{d}b}$ 4. $g(\theta) = -2\theta^{-2}\cos(3\theta)$; $g'(\theta)$

2. $y = 5x\cos(2x)$; $\dfrac{\mathrm{d}y}{\mathrm{d}x}$ 5. $a = -4t^{-1}\mathrm{e}^{-2t}$; $\dfrac{\mathrm{d}a}{\mathrm{d}t}$

3. $f(x) = x^2 \mathrm{e}^{3x}$; $f'(x)$ 6. $P(t) = (\sin(6t)) \times (5\ln(6t))$; $P'(t)$

You may find it helpful to identify the constants and separate these from the variables as it will make your differentiation simpler.

Identify the two functions for the product rule first; when differentiating these, you will then see the need for the chain rule.

? Exercise 7.11 The Maximum Value of the Maxwell–Boltzmann Distribution

Having practised both the chain rule and the product rule, use these to find the first derivative of the Maxwell–Boltzmann distribution and use it to identify an expression for the speed at which the distribution is a maximum. Remember that k_B and π are constants, and, for a given system, T and m are to be considered constant also:

$$f(v) = 4\pi \left(\frac{m}{2\pi k_\mathrm{B} T} \right)^{3/2} v^2 \mathrm{e}^{\left(-mv^2/2k_\mathrm{B}T \right)}$$

7.4.3 Quotient Rule

The quotient rule is not as widely used in chemistry, finding most applications in the area of quantum and computational chemistry. It is included here for the sake of completeness as you may find it useful, but it is also possible to solve problems of this structure by combined use of the chain and product rules.

❶ Key Concept 7.8 Quotient Rule—Differentiating Compound Functions

For a compound function of the form

$$y = \frac{u}{v}$$

where u and v are functions of x, the first derivative may be found by applying the *quotient rule*:

$$\frac{\mathrm{d}y}{\mathrm{d}x} = \frac{v\left(\dfrac{\mathrm{d}u}{\mathrm{d}x} \right) - u\left(\dfrac{\mathrm{d}v}{\mathrm{d}x} \right)}{v^2} \tag{7.39}$$

The quotient rule can be seen as an application of the product rule, as the quotient can be rewritten as a product:

$$y = \frac{u}{v} \equiv uv^{-1}$$

The product rule may then be applied to obtain result shown in eqn (7.39).

Example 7.8 Application of Quotient Rule to Trigonometric Functions

In this example, we will use the formula for the quotient rule (eqn (7.39)) and the known derivatives of the sine and cosine functions shown in Key Concept 7.5 to find the derivative of the tangent function, $\tan(Ax)$.

Recall that $\tan(Ax)$ has the following identity:

$$\tan(Ax) = \frac{\sin(Ax)}{\cos(Ax)}$$

We can now make the substitutions of u and v to set up our quotient:

$$u = \sin(Ax) \qquad \frac{du}{dx} = A\cos(Ax)$$

$$v = \cos(Ax) \qquad \frac{dv}{dx} = -A\sin(Ax)$$

We can now apply the formula for the quotient rule to find the first derivative of $\tan(Ax)$:

$$y = \tan(Ax) \equiv \frac{\sin(Ax)}{\cos(Ax)}$$

$$\frac{dy}{dx} = \frac{v\left(\dfrac{du}{dx}\right) - u\left(\dfrac{dv}{dx}\right)}{v^2} = \frac{\left(\cos(Ax) \times A\cos(Ax)\right) - \left(\sin(Ax) \times -A\sin(Ax)\right)}{\left(\cos(Ax)\right)^2}$$

$$= \frac{A\left(\cos^2(Ax) + \sin^2(Ax)\right)}{\cos^2(Ax)} \equiv \frac{A}{\cos^2(Ax)}$$

$$= A\sec^2(Ax) \qquad\qquad (7.40)$$

You will recall the trigonometric identity $\cos^2(Ax) + \sin^2(Ax) = 1$ from Chapter 5; this allows us to make the simplification to the final form in eqn (7.40).

7.5 Partial Differentiation of Multiple Variables

We are taught early on in our scientific training that, when conducting an experiment, we must '*only change one thing at a time*'; indeed, many of our models rely on this principle, calculus included. All of the examples we have seen so far in our studies of differentiation look at the rate of change of an observable property as another property varies; for example, examining the rate of change of reactant concentration as reaction time progresses (the rate of reaction, Section 7.1.1), or the rate of change of interatomic potential as the atoms in a diatomic molecule move closer together (the interatomic force, Section 7.1.2). The reality is that, for many reactions and processes, a number of variables can affect the observed outcome, and it is our experimental design which restricts the freedom of the system such that we only need consider a single variable outcome. When modelling molecules in 'real life', however, we must consider the multivariate nature of the models which we use.

7.5.1 Ideal Gas Considerations

Most students of chemistry will readily recall the ideal gas equation:

$$pV = nRT \tag{7.41}$$

The equation neatly ties together Boyle's law (pV = constant), Charles' law ($V \propto T$) and Avogadro's law ($V \propto n$). However, for each of these laws to be true, there is a requirement that *the other possible variables remain constant*. The result is that we have three interrelated variables which can all be plotted together on a three-dimensional plot, as shown in Figure 7.9.

A gradient of this three-dimensional surface will therefore be dependent on each of these three variables; mathematically, this becomes troublesome. We instead calculate a *partial derivative*, where we identify two correlated variables and we consider the variation of these *while keeping any other possible variables constant*.

In the context of the ideal gas equation (eqn (7.41)), we may wish to identify how the pressure varies with respect to temperature. To do this, we must fix the third variable, volume, and rewrite the gas law as:

$$p = \frac{nR}{V}T \tag{7.42}$$

To find the gradient of this graph, we now calculate a *partial derivative*, in which we apply the same technique for differentiating simple functions as before (Key Concept 7.2), but considering the volume V to be constant:

$$\left(\frac{\partial p}{\partial T}\right)_V = \frac{nR}{V} \tag{7.43}$$

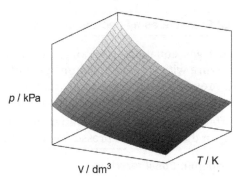

Figure 7.9   The interrelation of pressure (p), temperature (T) and volume (V) for a fixed quantity of an ideal gas. Note the linear relationship between T and p, but the non-linear relationship between V and p.

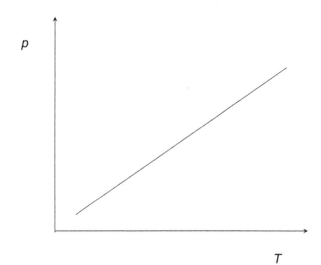

Figure 7.10   The relationship between pressure and temperature for an ideal gas at a constant volume.

We can 'translate' eqn (7.43) to say that '*the rate of change of pressure p with respect to temperature T while keeping volume V constant is equal to...*', and as this derivative is a constant (because we only consider it at constant volume), we see the expected straight-line relation shown in Figure 7.10. The 'curly d' shows that we are calculating a partial derivative, while the subscript V states what we are keeping constant (in this case the V is volume).

?Exercise 7.12 Pressure as a Function of Volume

- Using the ideal gas equation, rearrange and find the rate of change of pressure with respect to volume at constant temperature, $(\partial p/\partial V)_T$. You will need to consider what the constants are, and how to calculate a first derivative in the form $1/x$.
- How is this different from the first derivative shown in eqn (7.43)?
- Sketch a graph for the variation of pressure with respect to volume.

7.5.2 Thermodynamic Considerations

In thermodynamics, we have a wide array of variables to consider. The total internal energy of a system, U, is a result of the contributions from the overall entropy of the system, S, the volume of the system, V, and the composition of the system (number of molecules of different types in the system; n_A, n_B and so on).[‡] In the context of a chemical reaction, all of these values will be in constant flux, so anticipating the rate of change of the internal energy demands a partial derivative.

Depending on the partial derivative, we can determine a different physical property of the system; examples of this are:

$$T = \left(\frac{\partial U}{\partial S}\right)_{V,N} ; \quad p = -\left(\frac{\partial U}{\partial V}\right)_{S,N} \tag{7.44}$$

Thus the temperature of a system at constant volume and number of molecules is equal to the rate of change of the internal energy, U, with respect to entropy, S, while the pressure, p, of a system of fixed entropy and composition is found from the rate of change of the internal energy with respect to volume at constant S, N.

A complete discussion of these thermodynamic state functions is outwith the scope of this text, however this illustrates the power of the partial derivative in thermodynamics.

?Exercise 7.13 Partial Derivative of the Gibbs Energy

- From the general form of the Gibbs equation (eqn (7.45)), find an expression for the rate of change of the Gibbs energy with respect to temperature at constant entropy and enthalpy. Write this in the appropriate partial derivative notation

$$G = H - TS \tag{7.45}$$

- Sketch a graph of the variation of the Gibbs energy with temperature.

[‡]In a rigorous statistical model, the entropy, S, is the state of consideration, with the internal energy, U, being allowed to vary along with the volume and composition. Historically, U was the primary consideration as this was based on empirical observations of energy transfer. We have used this to illustrate the rate of change as it is more straightforward for the novice to identify the states described in eqn (7.44); however, this approach is now considered outdated.

7.5.3   Quantum Chemistry Considerations

We have already said that finding the 'rate of change' is necessary when considering a quantum mechanical wave function (Section 7.2); the curvature of the wave function (its second derivative) gives us information about the kinetic energy of a quantum mechanical particle. However, any given quantum mechanical particle is capable of moving in three dimensions along the x-, y- and z-axes. When considering many quantum mechanical particles, we usually wish to simplify this to consider the kinetic energy (E_k) in a single dimension and assume no movement in the other dimensions, shown as:

$$\left(\hat{T}_x \Psi\right) = -\frac{\hbar}{2m}\left(\frac{\partial^2 \Psi}{\partial x^2}\right)_{y,z} \qquad (7.46)$$

> The symbol $\hbar$ (known as "h-bar") is the reduced Planck constant, equal to $h/2\pi$.

? Exercise 7.14 Determining the Kinetic Energy of a Quantum Mechanical Particle

---

A general form of the wave function for a quantum mechanical particle moving in a restricted space is given in eqn (7.47).[§]

- Assume that the particle can only move in a single dimension; use the overall form of the wave function to determine an expression for its kinetic energy in this dimension

$$\Psi_{\{x,y,z\}} = A\sin(kx) + B\sin(ky) + C\sin(kz) \qquad (7.47)$$

- Sketch a graph of the wave function in this dimension.

7.6 Summary

In this chapter, we have looked at the importance of being able to find the gradient of a function; as we find mathematical models to explain the behaviour of chemical systems, we need to be able to apply mathematical techniques to make predictions about those systems. Finding turning points for chemical functions allows us to determine equilibrium positions of systems, maximum values of distributions and nodes in quantum mechanical wave functions. Some of the mathematical models we use to describe chemistry require simultaneous variation of many different variables; in using the technique of partial differentiation, we are able to treat these relationships using the same mathematical approach to find the rates of change.

[§]This is often called the *'particle in a box'* model.

Summary: Chapter 7

Differentiation is the process of finding the gradient of a function at any point; the key concepts covered in this chapter are listed here.

First Derivatives of Simple Functions

$f(x) \equiv y$	$f'(x) \equiv \dfrac{dy}{dx}$
A	0
Ax^n	nAx^{n-1}
$\dfrac{A}{x^n} \equiv Ax^{-n}$	$-nAx^{-n-1}$
e^{Ax}	Ae^{Ax}
$A\ln(Bx)$	$\dfrac{A}{x}$
$\sin(Ax)$	$A\cos(Ax)$
$\cos(Ax)$	$-A\sin(Ax)$
$\tan(Ax)$	$-A\sec^2(Ax)$
$Au + Bv$	$A\dfrac{du}{dx} + B\dfrac{dv}{dx}$

Note: This list is not exhaustive, and other functions may also be differentiated. Please note: A and B are constants, and u and v functions of x.

Chain Rule, Product Rule and Quotient Rule

In all the following examples, the terms u and v indicate different functions of x.

	$f(x) \equiv y$	$f'(x) \equiv \dfrac{dy}{dx}$
Chain rule	$y = f(u); \; u = g(x)$	$\dfrac{dy}{du} \times \dfrac{du}{dx}$
Product rule	uv	$u\dfrac{dv}{dx} \times v\dfrac{du}{dx}$
Quotient rule	$\dfrac{u}{v}$	$\dfrac{v\left(\dfrac{du}{dx}\right) - u\left(\dfrac{dv}{dx}\right)}{v^2}$

- Use the chain rule when one function acts the output of another, *i.e.* they are nested, *e.g.* $\sin(x^2)$.
- Use the product rule when the output of one function is multiplied by the output of another function, *e.g.* $x^2 \times \sin(x)$.
- Use the quotient rule if the output of one function is divided by the output of another and you cannot rewrite it as a product (the arithmetic is simpler with the product rule) *e.g.* $\sin(x)/x^2 \equiv x^{-2} \times \sin(x)$.

❓Exercise 7.15 Differentiation Problems in a Chemical Context

The following questions are each an example of how differentiation is used in chemistry; they combine different aspects of what has been covered earlier in this chapter, and in some examples you may need to combine techniques.

1. The Madelung equation may be used to determine the internal energy, ΔU, (J mol^{-1}) between two neighbouring atoms:

$$\Delta U = -\frac{N_A A |z_+||z_-| e^2}{4\pi \varepsilon_0 r} \qquad (7.48)$$

where $|z+|$ and $|z-|$ are the dimensionless charges (in C) on the anion and cation respectively, r is the separation of the ions (in m) and A is the Madelung constant. Determine the force of attraction between the two ions, $d\Delta U/dr$.

2. For the wave function of a $2p_x$ orbital, given by:

$$\Psi = N e^{-r} 2a_0 r e^{i\phi} \sin\theta$$

determine the partial derivatives of the function with respect to each of the radial and angular terms, r, ϕ and θ.

3. The general form of a wave function Ψ for a free particle travelling along the x-axis is given by:

$$\Psi = K \cos\left(\frac{2\pi x \sqrt{2mE}}{h}\right)$$

where E is the energy of the particle. Determine an expression for $d\Psi/dE$.

4. The virial equation is used as an improvement to the ideal gas equation:

$$pV_m = RT\left(1 + \frac{B}{V_m} + \frac{C}{V_m^2}\right)$$

Rearrange the equation and determine an expression for dp/dV_m.

5. Determine the momentum, p_x, of a wave–particle whose wave function is given by:

$$\Psi = A e^{2\pi i x/\lambda} e^{-2\pi i \nu t}$$

by differentiating the function with respect to x and multiplying the result by the constant $-ih/2\pi$.[*]

6. When ions travel in solution, their conductivity is given by an equation which contains a temperature-dependent constant, b, given by:

$$b = \frac{qz^3 \varepsilon F}{24\pi \varepsilon_0 RT}\left(\frac{2}{\varepsilon RT}\right)^{1/2}$$

---

[*]You can learn more about 'i' in Chapter 9.

where z is the charge on the ion, T is the absolute temperature and all other values are constants.

(a) Multiply out the bracket to give an expression for b with only one T term.

(b) Determine how the value of the constant b changes with temperature, db/dT.

7. The radius of a spherical balloon increases at a rate of $0.15\,\mathrm{m\,s^{-1}}$. Find the rate of increase of the volume with respect to time.

8. Non-linear effects in lasers occur such that the electric field of the incident light E follows the expression:

$$E = \sqrt{\frac{E_0^2\left(1+\cos(2\omega t)\right)}{2}}$$

where t is the time during the oscillation cycle and all other terms are constant. Write an expression for dE/dt.

9. Light is scattered as it passes through solutions containing small particles with an intensity I, given by

$$I = I_0\,\frac{\pi\alpha^2}{\varepsilon_r^2\lambda^4 r^2}\,\sin^2\phi$$

Determine expressions for how the intensity varies with:

(a) Wavelength, λ
(b) Distance from scatterer, r
(c) Angle, ϕ

10. The radial distribution function for a 1s electron, Ψ_{rdf}, is given by

$$\Psi_{\mathrm{rdf}} = 4\pi r^2 K e^{-r/a}$$

Determine an expression for the most probable distance from the nucleus of the electron by differentiating the radial distribution function with respect to the distance, r.

Solutions to Exercises

Solutions: Exercise 7.1

1. $\dfrac{da}{db} = 5$

2. $\dfrac{dx}{dz} = \dfrac{z^2}{2}$

3. $\dfrac{ds}{dt} = 2t + 2$

4. $\dfrac{dy}{dx} = 6x + 2$

5. $\dfrac{df}{dg} = 5g^3 + g$

6. $\dfrac{dp}{dm} = 6m^2 + 6$

7. $\dfrac{dn}{dp} = 2p^3 + \dfrac{p}{2}$

8. $\dfrac{dc}{db} = 2\left(3b^2 + 1\right) = 6b^2 + 2$

(sidebar box)

$\dfrac{dr}{dt} = 0.15\,\mathrm{ms^{-1}}$, and

$V = \dfrac{4\pi r^3}{3}$.

Solutions: Exercise 7.2

1. $f'(a) = 5$

2. $\dfrac{dg}{df} = 6f$

3. $f'(t) = 8 + 3t$

4. $f'(a) = 27a^8 - \dfrac{6}{a^3}$

5. $f''(b) = 36b^{-5} + 40b^3 - 360b^{-10}$

6. $\dfrac{dx}{dy} = 2y^{-1/2}$

7. $f'(x) = 6x^{-1/3} + x^{-2/3}$

8. $\dfrac{dz}{da} = 6.4a^{-0.2} + 2.4a^{-1.4}$

9. $\dfrac{dp}{dT} = \dfrac{nR}{V}$

10. $\dfrac{dG}{dT} = -S$

Solutions: Exercise 7.3

1. $a'(b) = 2e^{2b}$

2. $\dfrac{dy}{dx} = 6e^{2x} + 2$

3. $\dfrac{dr}{ds} = 2e^{2s} - 4e^{-2s}$

4. $p'(q) = -28e^{-7q} = -\dfrac{28}{e^{7q}}$

5. $\dfrac{df}{dg} = \dfrac{-e^{-g/4}}{8}$

6. $\dfrac{d[A]}{dt} = k[A]_0\, e^{-kt}$

Solutions: Exercise 7.4

1. $\dfrac{da}{db} = \dfrac{5}{b}$

2. $y'(x) = \dfrac{1}{x}$

3. $\dfrac{dr}{ds} = \dfrac{2}{s} - 2e^{-2s}$

4. $\dfrac{dp}{dq} = \dfrac{2}{q}$

5. $f'(g) = \dfrac{2}{g}$

6. $\dfrac{dy}{dz} = \dfrac{1}{z} + \dfrac{2}{z} = \dfrac{3}{z}$

Solutions: Exercise 7.5

1. $\dfrac{da}{db} = 15e^{3b}$

2. $\dfrac{dy}{dx} = 5\cos x$

3. $f'(x) = 36e^{2x} + \dfrac{2}{x}$

4. $r'(s) = -2s^{-3} - 0.5s^{-1.5}$

5. $\dfrac{dz}{dy} = -a^2 e^{-ay} + b^2 e^{by}$

6. $g'(a) = -9\cos(3a) - 8\sin(2a)$

7. $\dfrac{d\phi}{d\theta} = -8k\,\sin(k\theta)$

8. $\dfrac{d\Psi}{dx} = Abe^{bx}$

9. $V'(r) = \dfrac{-q_1 q_2}{4\pi\,\varepsilon_0 r^2}$

10. $\dfrac{dV(r)}{dr} = 4\varepsilon\left(\dfrac{-12r_0^{12}}{r^{13}} + \dfrac{6r_0^6}{r^7}\right)$

In our solutions we have used Lagrange notation to show the derivatives as it is less cluttered than the Leibniz's notation (see Table 7.1 for more information), however it still conveys the same information.

Solutions: Exercise 7.6

1. $f'(b) = 10b + 6$ Turning point: $a = -1.8$, $b = -0.6$

2. $f'(h) = 6h - 2$ Turning point: $g = \dfrac{-1}{3}$, $h = \dfrac{1}{3}$

3. $f'(a) = 9a^2 + 6a$ Turning point 1: $p = 0$, $a = 0$
   Turning point 2: $p = \dfrac{4}{9}$, $a = -\dfrac{2}{3}$

4. $f'(x) = -64x^{-3} + 8$ Turning point:[††] $x = 2$, $y = 24$

5. $f'(h) = 3h^2 - 3$ Turning point 1: $g = 0$, $h = 1$;
   Turning point 2: $g = 4$, $h = -1$

6. $f'(m) = 36m^2 - 24m + 4 \equiv 4(9m^2 - 6m + 1) \equiv 4(3m - 1)^2$
   Turning point: $m = \dfrac{1}{3}$, $l = \dfrac{-32}{9}$

Solutions: Exercise 7.7

1. $f''(b) = 10$ Turning point: $f''(b) > 0 \rightarrow$ minimum

2. $f''(h) = 6$ Turning point: $f''(h) > 0 \rightarrow$ minimum

3. $f''(a) = 18a + 6$ Turning point 1: $f''(a) > 0 \rightarrow$ minimum
   Turning point 2: $f''(a) < 0 \rightarrow$ maximum

4. $f''(x) = 192x^{-4}$ Turning point: $f''(x) > 0 \rightarrow$ minimum

5. $f''(h) = 6h$ Turning point 1: $f''(h) > 0 \rightarrow$ minimum;
   Turning point 2: $f''(h) < 0 \rightarrow$ maximum

6. $f''(m) = 72m - 24$ Turning point: $f''(m) = 0 \rightarrow$ inflection

Solutions: Exercise 7.8

1. $f'(a) = 2a\cos(a^2)$

2. $\dfrac{dy}{dx} = -60x^2\sin(5x^3)$

3. $f'(b) = 5e^{\sin(5b)}\cos(5b)$

4. $\dfrac{dc}{db} = -\left(2b - 3b^{-2}\right)\cos\left(b^2 + 3b^{-1}\right)$

5. $\dfrac{dy}{dx} = \dfrac{-10e^{5/x^2}}{x^3}$

6. $f'(v) = \dfrac{3v^2 e^{v^3/2}}{2}$

[††]Formally, there are other turning points; see Section 9.2.1.

Solutions: Exercise 7.9

$$V'_{\text{Morse}}(r) = 2Da\ e^{-a(r-r_{\text{eq}})} \times \left(1 - e^{-a(r-r_{\text{eq}})}\right)$$

Solutions: Exercise 7.10

1. $\dfrac{da}{db} = \sin(2b) + 2b\cos(2b)$

2. $\dfrac{dy}{dx} = 5\cos(2x) - 10x\sin(2x)$

3. $f'(x) = 2xe^{3x} + 3x^2 e^{3x} \equiv xe^{3x}(2 + 3x)$

4. $g'(\theta) = 4\theta^{-3}\cos(3\theta) + 6\theta^{-2}\sin(3\theta) \equiv 2\theta^{-2}(2\theta)^{-1}\cos(3\theta) + 3\sin(3\theta))$

5. $\dfrac{da}{dt} = 4t^{-2}e^{-2t} + 8t^{-1}e^{-2t} \equiv 4t^{-1}e^{-2t}(t^{-1} + 2)$

6. $P'(t) = (6\cos(6t))(5\ln(6t)) + 5t^{-1}\sin(6t)$

Solutions: Exercise 7.11

$$f'(v) = 8\pi\left(\frac{m}{2\pi k_B T}\right)^{\frac{3}{2}} ve^{-\left(\frac{m}{2k_B T}\right)v^2}\left(1 - \left(\frac{m}{2k_B T}\right)v^2\right)$$

$$v_{\text{most probable}} = \sqrt{\frac{2k_B T}{m}}$$

Solutions: Exercise 7.12

- $\left(\dfrac{\partial p}{\partial V}\right)_T = -\dfrac{nRT}{V^2}$

- The variation of pressure with respect to temperature is independent of temperature (the gradient is constant), while for the variation of pressure with respect to volume we find that the function still depends on the volume.

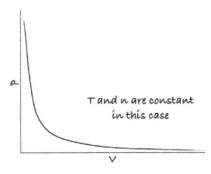

T and n are constant in this case

As in Exercise 4.2, you are asked for a sketch; only indicate the shape of the graph and the key characteristics (any intersections, gradients etc.).

Solutions: Exercise 7.13

• $\dfrac{dG}{dT} = -S = -S$

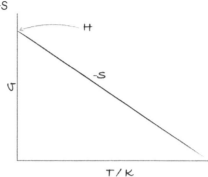

Solutions: Exercise 7.14

• $E_K = \dfrac{Ak^2\hbar}{2m}\sin(kx)$

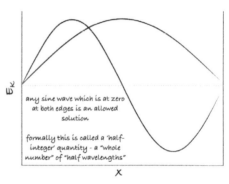

any sine wave which is at zero
at both edges is an allowed
solution

formally this is called a 'half-
integer' quantity - a "whole
number" of "half wavelengths"

Solutions: Exercise 7.15

1. Solve using direct differentiation;

$$F = \frac{d\Delta U}{dr} = \frac{N_A A|z_+||z_-|e^2}{4\pi\varepsilon_0 r^2}$$

2. a) Solve using product rule; $\left(\dfrac{\partial\psi}{\partial r}\right)_{\theta,\phi}$

$$= -Ne^{-r}2a_0re^{i\phi}\sin\theta + Ne^{-r}2a_0e^{i\phi}\sin\theta \equiv 2Ne^{i\phi}e^{-r}\sin\theta(1-r)$$

   b) Solve using direct differentiation; $\left(\dfrac{\partial\psi}{\partial\theta}\right)_{r,\phi} = Ne^{-r}2a_0re^{i\phi}\cos\theta \equiv 2Na_0re^{-r}e^{i\phi}\cos\theta$

   c) Solve using direct differentiation; $\left(\dfrac{\partial\psi}{\partial\phi}\right)_{r,\theta} = iNe^{-r}2a_0re^{i\phi}\sin\theta \equiv 2iNa_0re^{-r}e^{i\phi}\sin\theta$

3. $\left(\dfrac{\partial\psi}{\partial E}\right)_x = -\dfrac{K2\pi x}{2}\sqrt{\dfrac{2m}{E}}\cdot\sin\left(\dfrac{2\pi x\sqrt{2mE}}{h}\right)$

4. Solve using direct differentiation; $\dfrac{dp}{dV_m} = -\dfrac{RT}{V_m^2}\left(1 + \dfrac{2B}{V_m} + \dfrac{3C}{V_m^2}\right)$

5. $p_x = -\dfrac{Ai^2 h}{\lambda}e^{\frac{2\pi\, ix}{\lambda}}e^{-2\pi\, i\nu t}$

> The i in this question is an imaginary term where $i^2 = -1$, you can learn more about this in Chapter 9.

6. (a) Part a: $b = \dfrac{qz^3 F}{24\pi\varepsilon_0}\cdot\sqrt{\dfrac{2\varepsilon}{R^3 T^3}} \equiv \dfrac{qz^3 F}{24\pi\varepsilon_0}\cdot\sqrt{\dfrac{2\varepsilon}{R^3}}\cdot T^{-3/2}$

(b) Part b, solve using direct differentiation : $\dfrac{db}{dT} = -\dfrac{qz^3 F}{16\pi\varepsilon_0}\cdot\sqrt{\dfrac{2\varepsilon}{R^3}}\cdot T^{-5/2} \equiv \dfrac{qz^3 F}{24\pi\varepsilon_0}\cdot\sqrt{\dfrac{2\varepsilon}{R^3 T^5}}$

7. Solve using chain rule; $\dfrac{dV}{dt} = 0.6\pi r^2$

8. Solve using chain rule $\dfrac{dE}{dt} = -\omega\sin(2\omega t)\sqrt{\dfrac{E_0^2}{2\left[1 + \cos(2\omega t)\right]}}$

9. (a) Solve directly : $\left(\dfrac{\partial I}{\partial\lambda}\right)_{r,\phi} = -4I_0\,\dfrac{\pi\alpha^2}{\varepsilon_r^2\lambda^5 r^2}\sin^2\phi$

(b) Solve directly : $\left(\dfrac{\partial I}{\partial r}\right)_{\lambda,\phi} = -2I_0\,\dfrac{\pi\alpha^2}{\varepsilon_r^2\lambda^4 r^3}\sin^2\phi$

(c) Solve using chain rule : $\left(\dfrac{\partial I}{\partial\phi}\right)_{\lambda,r} = 2I_0\,\dfrac{\pi\alpha^2}{\varepsilon_r^2\lambda^4 r^2}\sin\phi\,\cos\phi$

10. Solve using product rule; $r = 2a$

Learning Points: What We'll Cover

- ☐ Introducing the 'antiderivative' as the inverse function for differentiation seen in Chapter 7
- ☐ Applying the 'antiderivative' to integration
- ☐ Handling the constant of integration and recognising its origin
- ☐ Applying integration to chemical kinetics and quantifying the constant of integration to obtain a rate law
- ☐ Eliminating the constant of integration by using definite integration between two limits
- ☐ Using definite integrals to calculate thermodynamic work and reintroducing the principles of calculus to calculate areas under curves
- ☐ Advanced integration techniques for compound functions, through 'integration by substitution' and 'integration by parts'
- ☐ Introducing tables of standard integrals and how to use them in calculations

Why This Chapter Is Important

- The area enclosed by a function when plotted on a graph can be used to quantify a range of chemical changes, including thermodynamic work, entropy, *etc.*
- Integration is the process of finding the area and is the inverse process to differentiation, called an 'antiderivative'.
- Functions which involve a rate of change can be integrated to find versions of equations that can model and predict chemical processes, *e.g.* rates of reactions.

Calculus 2, Integration: Reaction Kinetics and Rate Laws

In Chapter 7, we explored the meaning of gradients in plotting relationships in chemical phenomena. We now consider a further aspect of chemical change, that which happens when we *multiply* two variables together under changing conditions—in other words, finding the area enclosed by a curve. This allows us to calculate the thermodynamic work from a pressure–volume graph, the potential energy change from a force–distance graph or a proportion of a population from a distribution of molecular speeds.

In this chapter, we introduce the principle of the antiderivative and show how it can be used to calculate areas exactly using the principles of calculus first introduced in Chapter 7, namely 'infinitesimal change' and summing 'in the limit'. We also show how a rate law can be constructed using the same principles if we start with the concept of the rate of a reaction. Finally, we examine how to integrate more complicated functions using integration processes analogous to the 'chain rule' and 'product rule' introduced in Chapter 7.

8.1 The Need for an Inverse Function

Throughout this book, we have always considered our mathematical operations as complementary pairs—a pair of functions, each of which will 'undo' the work of the other. This is how we considered addition and subtraction, multiplication and division, exponents and logarithms, *etc*. We can now apply this logic to our treatment of calculus; after all, whatever mathematical operation we perform, we have to be able to undo it. We explored the process of differentiation in Chapter 7 to find the gradient of a curve; can we instead find the equation of a curve if we know how its gradient varies? It turns out that we can, but with certain limitations.

Often in chemistry we measure a *rate of change* directly; the main example being 'the rate of reaction'. To build our chemical model in this example, we need to get back to the original question, such as 'How does the concentration of reactant change with time?' Since we are already starting with the 'rate of change', we need to 'undo' the derivative; *i.e.* to use the inverse function to return to the original statement of how the concentration varies with time.

We will firstly look at the harmonic oscillator example, as this involves relatively simple mathematics to demonstrate principles of integration, and then move on to other examples in chemistry, as well as other applications of integration.

8.2 Integration—Exploring the Harmonic Oscillator

The example shown in Section 7.1.2 showed the potential energy of a harmonic oscillator and identified that the gradient of this curve would give us the magnitude and direction of the restoring force on the two atoms in the vibrating diatomic molecule. However, we would not normally be able to measure the stored potential energy directly. Instead, we need to carry out an experiment in which we measure how the force on the atoms varies as we change the intermolecular distance. A similar experiment that we could do on the bench would involve two masses on a spring; we would measure the force between the masses as the spring is stretched (Figure 8.1).

The units of energy are joules ($J = kg\ m^2\ s^{-2}$ and the units of force are newtons ($kg\ m\ s^{-2}$).

Through consideration of units (Chapter 1), we know that the force is a rate of change of energy with respect to distance; *i.e.* force is the first derivative of the potential in the spring. Therefore, to get back to the statement of how potential varies, we need to 'undo'

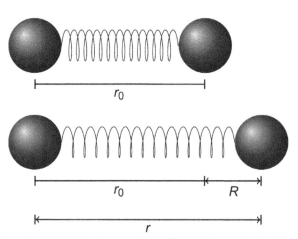

Figure 8.1   Revisiting the harmonic oscillator. The restoring force in the bond is proportional to the extension (or compression), R, from the equilibrium separation, r_0. From this restoring force, it is possible to identify the potential energy stored in the bond using the principles of integration.

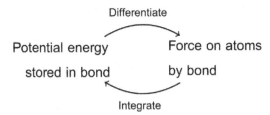

Figure 8.2 Through applying differentiation or integration with respect to interatomic separation we can move between expressions for potential energy between and the force on atoms in a bond.

the differentiation in a process called *integration* (*i.e.* we will *integrate* the force). This allows us to work backwards and determine an expression for the potential (Figure 8.2).

If we start with the mathematical model describing the force, can we reconstruct the graph of the potential energy? In other words, can we reconstruct the original curve if we are only told how the gradient varies? Let us work through the steps we need to rebuild the original equation.

1. Remember the equation for the force, F as a function of the bond extension R in a harmonic oscillator (Figure 8.1):

$$F = -kR \qquad (8.1)$$

2. Since R is the *extension* of the bond from its equilibrium bond length, we can replace this as $R = r - r_0$ to give the equation solely in terms of the absolute distance between atoms, r:

$$F = -k(r - r_0) \equiv -kr + kr_0 \qquad (8.2)$$

3. Now we ask ourselves, '*What expression gives us this result when differentiated?*' Fortunately, we already know what we are aiming at (see Example 7.2); however, let's just consider what needs to happen for any function for which we wish to find the integral. Consider the general function Ax^n:
   - When we differentiate a function, the index of the variable (power) *decreases* by one (*i.e.* we subtract the value of 'one' from the index, $n - 1$); therefore, to go the other way (to integrate), the index of the variable will *increase* by one (*i.e.* we add the value of 'one' to the index, $n + 1$; see Key Concept 7.2).
   - Additionally, when the function is differentiated, the result is *multiplied* by the value of its *original* index (nA); therefore, for integration the resulting function is *divided* by the value of its *final* index $(A/(n+1))$.
   - Let's look at this in algebraic terms. We are already familiar with finding the derivative, but now let's look at the integral (Figure 8.3).

The negative value is there because it is a *restoring force*; a positive change in position results in a force pushing back towards the starting point; likewise, a negative change in position results in a force pushing in the positive direction towards the starting point.

Remember that k is the force constant of the spring.

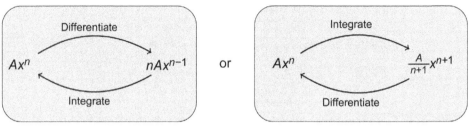

Figure 8.3   On the differentiation path, 'one' is subtracted from the value of the index, while on the integration path 'one' is added to the value of the index. In the first cycle the rules of differentiation are applied first, then reversed to show 'integration'. In the second cycle, we have applied the rules of integration first, then differentiated to get back to the starting point.

Returning to the harmonic oscillator, we can now determine the integral for this expression. As eqn (8.2) is simply a summation of polynomials (*i.e.* simple powers of the variable are added together), we can integrate each term separately.

4. For $(-kr)$:
   - We must add 'one' to the value of the index of r; this means that the power of r goes from one to two: r^2.
   - We must also divide the coefficient by the final power, so $-k$ becomes $-k/2$.

   - Our final integral for $-kr$ is therefore $-kr^2/2$.

   - We can check that this is correct by differentiating $-kr^2/2$, and we indeed return to where we started.
5. For $+kr_0$, it does not appear that we have any powers of r present; however, as this is a constant, there is an invisible $r^0 \equiv 1$; we can therefore write this as $+kr_0r^0$:
   - The power of r goes from zero to one: $r^1 \equiv r$.
   - For the coefficient, we now need to divide this by the final power, one. However, this will not affect the coefficient so it remains unchanged.
   - Our final integral for $+kr_0$ is kr_0r; again, we can check this by differentiating the final expression to get back to our initial expression.
6. There is one last item that we need to be aware of; when we differentiated a function, any *constant* values did not change

Remember that r_0 is a *constant* denoting the equilibrium separation.

with respect to the variable, so these differentiated to zero. As such, we cannot know if there are any constants in the integrated function. For this reason, we add a *constant of integration*, + C, to our result.

> C is the constant of integration; see Key Concept 8.2.

7. Our final overall expression for the integral of the force equation for the harmonic oscillator is:

$$\int F \, dr = \int \left(-k(r - r_0) \right) dr = -\frac{kr^2}{2} + kr_0 r + C \qquad (8.3)$$

As ever, it can be helpful to translate the mathematics into plain English. The first component of eqn (8.3) becomes, 'The integral of the force, F, with respect to r is equal to …'. Notice, however, that we cannot recover the constant value of $kr_0^2/2$ which appears in eqn 7.5. The reason for this is shown in Key Concept 8.2.

We now have our method for finding the integral for a general function, and this is summarised in Key Concept 8.1.

---

❶ Key Concept 8.1 Integration: Finding the 'Antiderivative'

The mathematical name for the process of finding the 'antiderivative' is *integration*. If we have a function, $f(x)$, we use the following notation to describe the process:

Function $f(x) = Ax^n$

Integral ('anti-derivative') $\int f(x) \, dx = \int \left(Ax^n \right) dx = \frac{A}{n+1} x^{n+1} + C$

Notice that we have added a 'constant of integration', + C (see Key Concept 8.2). The 'integral' notation may also be described in plain English:

$$\int f(x) \, dx = \ldots$$

'The integral ($\int$) of a function of x ($f(x)$) with respect to x (dx) is equal to…'

Using this general form, we have identified the inverse function of differentiation, and named it *integration*.

> ❶ **Key Concept 8.2 The Constants of Integration**
>
> When we differentiate, we follow a rule whereby we subtract one from the value of the index and multiply the result by the value of the index; thus:
>
> $$f(x) = 3x^2 + 4x + 5$$
> $$\equiv 3x^2 + 4x^1 + 5x^0$$
> $$f'(x) = (2 \times 3)x + (1 \times 4)x^0 + (0 \times 5)x^{-1}$$
> $$\equiv 6x + 4$$
>
> However, if we try to work backwards, while we can reconstruct the $3x^2$ and $4x$ component of the original equation, we cannot recover the constant value 5. We have lost information in the process of differentiation.
>
> As we have lost information about the constant, we instead use a general constant to stand in; 'plus C':
>
> $$f'(x) = 6x + 4$$
> $$\int f'(x) = 3x^2 + 4x + C$$

By convention, this constant is usually shown in upper case.

Differentiating a simple power of a variable (a polynomial) is done in two steps; firstly, multiply by the index and then subtract *one* from the index to get the new index. Integration is the reverse of this workflow: we add one to the index and then we divide the whole value by the new index.

Notice that we have also reversed the order of steps. When differentiating, we multiply first, but when integrating, we apply the inverse function (division) last.

The same, *reverse workflow*, approach works for our other basic functions and so the integral of $\cos(x)$ with respect to x is $\sin(x) + C$ (see Table 8.1).

Table 8.1   A list of some antiderivatives used for integration. Note that some of these have conditions attached to them, as the antiderivative will not evaluate otherwise. A table with more antiderivatives is included in the summary at the end of the chapter.

$f'(x) \equiv y$	$f(x) \equiv \int y\, dx$	Condition
A	$x + C$	
Ax^n	$\dfrac{A}{n+1}x^{n+1} + C$	$x \neq -1$
$\dfrac{A}{x} \equiv Ax^{-1}$	$A\ln x + C$	
e^{Ax}	$\dfrac{1}{A}e^{Ax} + C$	
$\sin Ax$	$-\dfrac{1}{A}\cos Ax + C$	
$\cos Ax$	$\dfrac{1}{A}\sin Ax + C$	

? Exercise 8.1 Integration as the Inverse Function of Differentiation; Determining the 'Antiderivatives'

Using the table of 'antiderivatives' (Table 8.1), determine the 'antiderivative' of the following functions (always remember to include the $+ C$, the constant of integration).

1. $f'(x) = x^2 + x + 7$

2. $f'(p) = \dfrac{-3}{p^4}$

3. $\dfrac{da}{db} = b^{-3} + b^{-1/2}$

4. $\dfrac{dr}{ds} = s^{2/3} + 3s^{-2} + 2s + a$

5. $\dfrac{dn}{dm} = \dfrac{\cos(m)}{2}$

6. $\dfrac{dy}{dz} = -\cos\left(\dfrac{z}{4}\right) + 6z$

7. $f'(g) = \dfrac{4}{e^{ag}}$

8. $H'(T) = a + bT + cT^2$

9. $\dfrac{dS}{dT} = aT^2$

10. $V'(r) = -\dfrac{q_1 q_2}{4\pi\varepsilon_0 r^2}$

> If a letter appears and we are not integrating with respect to that letter, it can be considered a constant, so here the symbol a is just like any other constant.

Example 8.1 Identifying the Constant of Integration

If we know that our gradient at a given value is given by the formula '2x – 10', we can calculate that, at the point x = 6, we can substitute this value into the equation and find that the gradient will be equal to 2 *units*. If we need the general expression for our curve, we can integrate this expression according to the principles in Key Concept 8.2:

$$\int 2x - 10 \, dx = x^2 - 10x + C$$

However, whether the value of '+ C' is 25, 30, 35, or any other value, the gradients of all the possible curves will still be 2 units:

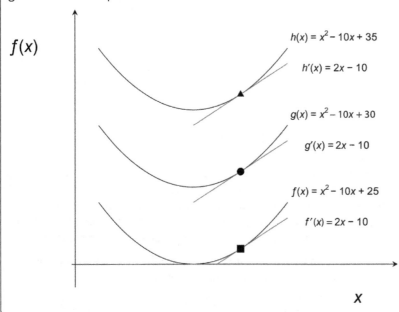

$f(x)$

$h(x) = x^2 - 10x + 35$

$h'(x) = 2x - 10$

$g(x) = x^2 - 10x + 30$

$g'(x) = 2x - 10$

$f(x) = x^2 - 10x + 25$

$f'(x) = 2x - 10$

X

With only an expression for the slope of the graph, $2x - 10$, it is impossible for us to know which of the graphs the gradient came from, as they each only differ by the constant. This is an illustration of the constant of integration and is a vital part of all indefinite integrals. If we know the coordinates of any single point through which the curve passes however, we can identify the constant.

For example, the curve $f(x) = 3x^2 + 4x + C$ is known to pass through the coordinate (8, 1). We can use these values ($f(x) = 8$ when $x = 1$) in the equation to identify C and hence to find the full equation for $f(x)$

$$f(x) = 3x^2 + 4x + C$$
$$8 = 3 \times (1)^2 + 4 \times (1) + C$$
$$= 3 + 4 + C$$
$$= 7 + C$$
$$C = 1$$
$$f(x) = 3x^2 + 4x + 1$$

？Exercise 8.2 Determining the Constants of Integration

Perform the integration required in each case, and determine values of the constant of integration, C in each case.

1. $y = \int 2x^2 \, dx$; when $x = 0$, $y = 5$

2. $f'(r) = \sin\left(\dfrac{r}{2}\right)$; when $r = \pi$, $f(r) = 2$

3. $p'(q) = e^{3q} + \dfrac{q}{2}$; when $q = 0$, $p(q) = 1$

4. $\dfrac{da}{db} = 3\cos 2b$; when $b = \pi$, $a = 3$

5. $z = \int w^3 + w^{\frac{1}{2}} + w^{-\frac{1}{2}} + 2a \, dw$; when $w = 0$, $z = 4a$

6. $\dfrac{dm}{dn} = \sin(2n) + \cos\left(\dfrac{n}{2}\right) + e^{2n}$; when $n = 0$, $m = 0$

8.3 Chemical Kinetics—Determining the Rate Law

In Section 7.1.1, we saw how a rate of reaction could be determined experimentally by determining the gradient of a concentration–time graph. However, what is less clear from this example is *how do we know which curve fits the data*? The process of identifying the rate at a given time allows us to set up a rate equation. Let's now revisit an experiment where we have hypothesised a second-order rate.

Through consideration of reaction mechanisms, we have hypothesised that this might be a unimolecular second-order reaction;

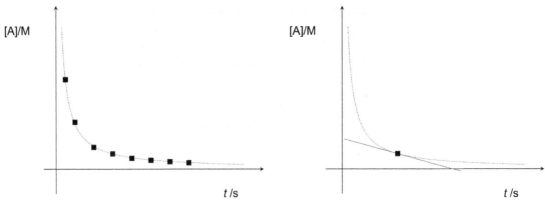

(a) Example concentration data from a second-order kinetics experiment.

(b) Determining the rate of reaction by calculating a gradient.

Figure 8.4 Example concentration data from a second-order kinetics experiment. Samples are taken at regular intervals and the concentration of the reactant A is measured. To determine the curve to fit the changing concentration, we need to further explore the rate law.

we therefore propose a rate equation based on the concentration of our reactant, [A]:

$$\text{rate} = \frac{d[A]}{dt} = -k[A]^2 \tag{8.4}$$

Remember that this is describing how the rate of reaction we measure relates to the concentration of the reactant A. However, to fit our data appropriately to determine whether our reaction is indeed second-order, we need to integrate the expression in eqn (8.4) to find how [A] varies with time.

Looking at the equation, it initially appears to be straightforward:

$$\frac{d[A]}{dt} = -k[A]^2 \tag{8.5}$$

However, remember that dA/dt is a rate of change of [A] *with respect to time*. Therefore, our integration needs to take place *with respect to time*:

$$[A] = -\int k[A]^2\, dt \tag{8.6}$$

This immediately presents us with a problem. The variable on the right-hand side is [A], *not t*; therefore we cannot calculate the integration in this manner. Instead, we can use a mathematical approach called *separating the variables* to group common variables together.

8.3.1 *Integrated Rate Law: Separating the Variables*

The principle of separating the variables in the context of eqn (8.5) is straightforward; firstly, identify your two variables (in this case [A] and t), collecting all of the [A] terms on one side of the equation and collecting all of the t terms on the other side. Formally, this is a bit of mathematical trickery, but as chemists we just need to know that it works. So let's now rearrange eqn (8.5), grouping the [A] terms on the left and the t terms on the right:

$$\frac{1}{[A]^2}d[A]=-k\,dt \qquad (8.7)$$

This looks slightly odd, as we have split up the derivative; something which is not really a fraction. How then do we integrate this? It turns out that this is a mathematical device which allows us to change our integration approach; instead of 'integrating both sides with respect to t' (as we tried to do in eqn (8.6)), we now integrate each side with respect to its labelled variable:

$$\int\frac{1}{[A]^2}d[A]=-\int k\,dt \qquad (8.8)$$

We now integrate the left side of eqn (8.8) with respect to [A] and we integrate the right side with respect to t. It may be easier to rewrite the fraction $\dfrac{1}{[A]^2}$ as a negative index, $[A]^{-2}$:

$$\int[A]^{-2}d[A]=-\int k\,dt \qquad (8.9)$$

We can now apply the integration rules introduced in Key Concept 8.1:

- For $\int[A]^{-2}d[A]$, we add one to the index, and divide by the final index, yielding $-[A]^{-1}$:

$$\int[A]^{-2}d[A]=\frac{[A]^{(-2+1)}}{(-2+1)}=-[A]^{-1}+C \qquad (8.10)$$

- For $-\int k\,dt$, there is the 'invisible' t^0; once again, add one to the index and divide by the final index, giving $-kt$:

$$-\int k\,dt\equiv-\int kt^0\,dt=-\frac{kt^{(0+1)}}{(0+1)}=-kt+C \qquad (8.11)$$

- We must remember that each of these has its own constant of integration; we can combine the constants of each integration into a single constant for the whole equation, D.

Our overall expression for the integral is now:

$$-[A]^{-1} = -kt - D$$

or:

$$\frac{1}{[A]} = kt + D \tag{8.12}$$

As illustrated in Key Concept 8.2, the general solution shown in eqn (8.12) can still apply to many equations, varying only according to the value of constant D. If we can identify a single point on the graph, we can gain a specific solution. Fortunately, if we are setting up a reaction, we have a known point on the graph—we know that at the start of the reaction ($t = 0$ s), we have a particular concentration of our reagent, the initial concentration $[A]_0$. This now allows us to find a value for D:

$$\text{At} \quad t = 0 \qquad [A] = [A]_0$$

$$\frac{1}{[A]_0} = (k \times 0) + D = D \tag{8.13}$$

$$\text{Therefore:} \qquad \frac{1}{[A]} = kt + \frac{1}{[A]_0}$$

We can now use this curve to fit the data; the satisfactory fit to the data in Figure 8.4a suggests that our prediction of a second-order rate for our reaction is correct. The equation may then be rearranged to make [A] the subject of the equation with respect to time and $[A]_0$.

> The constant of integration is needed on both sides of this equation; the two '+ C' terms have been combined into a single constant of integration, '+ D'.

8.3.2   First-order Rates: A Special Case

A first-order reaction needs to be treated slightly differently; however, we can go through the same process.

- Write the rate equation:

$$\frac{d[A]}{dt} = -k[A]$$

- Separate the variables:

$$\frac{1}{[A]} d[A] = -k \, dt$$

- Now, we set up the integration:

$$\int [A]^{-1} d[A] = -\int k \, dt$$

- While we can integrate the right-hand side with little problem, when we apply the integration procedure to the left-hand side we run into a problem:

$$\int [A]^{-1}\,d[A] = \frac{[A]^{-1+1}}{-1+1} + C$$

We cannot divide by zero. Fortunately, we already have a solution to this problem. In Key Concept 7.4, we saw how to differentiate the function ln(x), which is the complementary technique needed to carry out this integral, listed in Table 8.1.

- Once again, all constants of integration are combined into a single constant D:

$$\int [A]^{-1}\,d[A] = \ln[A] = -kt + D$$

- With our initial concentration $[A]_0$ at time $t = 0$, we can show that our integrated first-order rate law is:

$$\ln[A] = -kt + \ln[A]_0$$

We now have everything in place to determine an integrated rate law for any order of reaction.

---

❶ Key Concept 8.3 Integrating $1/x$

We cannot apply the standard integration formula to $1/x$ (or $1/t$, $1/v$, etc.), as we would end up dividing by zero.

Therefore, the following identity must be used:

$$\int \frac{1}{x}\,dx \equiv \int x^{-1}\,dx = \ln x + C$$

We can take the concept further with any scaling factors, as these can be taken outside the integral

$$\int \frac{1}{2V}\,dV \equiv \frac{1}{2}\int V^{-1}\,dV = \frac{1}{2}\ln V + C$$

This links to Key Concept 7.4 where the function ln x was differentiated to give $1/x$

❓Exercise 8.3 Integrating functions to natural logarithms

Perform the integration required in each case, and where possible determine values of the constant of integration, C.

1. $y = \int \left(\frac{2}{x}\right) dx$

2. $f'(r) = 1.5r^2 + r^{-1}$; when $r = 1$, $f(r) = 0$

3. $p'(q) = e^{q/2} + \dfrac{1}{2q}$

4. $z = \int w^{-1} + w^{1/2} + w^{-1/2}\,dw$; when $w = 1$, $z = 5$

5. $\dfrac{db}{da} = \dfrac{1}{a} + \ln 7$

6. $\dfrac{dm}{dx} = \dfrac{4}{x} + \dfrac{2}{x} + x$

8.4 Calculating Thermodynamic Work

We saw in Chapter 7 that we could gain additional results from an experiment by finding the gradient of a graph at a certain point, whether it was the rate of a reaction or the restoring force in a vibrating diatomic molecule. We also saw how we could determine the equilibrium separation for a diatomic molecule or the most probable speed of a gas molecule under Maxwell–Boltzmann kinetics by finding the point at which the gradient was equal to zero. All of this looked at how one variable changed with respect to another and, when we looked at the units, we saw that a gradient is found by *dividing* one variable by another. We must now look at what happens when variables are *multiplied* together. Take a look at the pressure–volume curve in Figure 8.5.

In Figure 8.5, an area is enclosed under the curve as the ideal gas is allowed to expand from volume V_1 to V_2. We know from our analysis of the units that when we multiply pressure and volume, pV, we obtain a unit of energy, and indeed the area represented by A in Figure 8.5 corresponds to the change in thermodynamic work when the ideal gas changes its volume in a reversible manner. Our challenge is to *determine the value of the enclosed area, A, for this* plot. We can then apply the same principles to determine the area enclosed by a curve for any chemical system; resolution of the units allows us to determine what the area means.

> The units of pressure are pascals (Pa = kg m^{-1} s^{-2}), and the units of energy are joules (J = kg m^2 s^{-2}).

While this problem may not immediately look like a 'rate of change', the area under the curve still changes with respect to

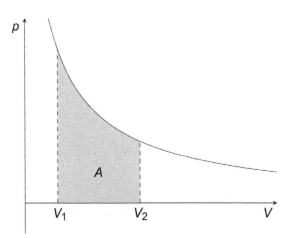

Figure 8.5 The relationship between pressure and volume for an ideal gas at a constant temperature. We are now considering the area, A, enclosed by the graph, which here indicates the thermodynamic work of a reversible change when the volume is changed from V_1 to V_2.

the independent variable[†]—in the case in Figure 8.5, the area A increases as V increases, but its rate of change slows as V increases— as such, we can consider it by using very similar principles.

8.4.1 Finding the Area Under a Curve

Let's continue our investigation into finding the work done when a gas is allowed to expand. For the purposes of this discussion, we are considering the gas to be expanding under *reversible* conditions.[‡] Under reversible conditions, the work of expansion (or compression) can be determined if we can find the area enclosed by our p–V graph. It is difficult to determine an area of a shape with a curved edge by inspection, so consider a rectangle of the appropriate dimensions instead (Figure 8.6).

The area of the rectangle A_1 shown in Figure 8.6 can be calculated as $A_1 = p_1 \Delta V$, however there is clearly a large error between the actual area under the curve and the value obtained. We can reduce the error here by finding a pressure at a volume midway between V_1 and V_2 (Figure 8.7).

Again, however, we still have an appreciable error in the end result. We can subdivide our rectangles still further, using a constant separation ΔV (Figure 8.8).

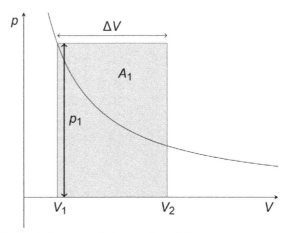

Figure 8.6   To find an area between V_1 and V_2, we can use a rectangle to start to approximate our determinations, with area $A_1 = p_1 \Delta V$. This is fairly crude and a large error is clearly apparent.

[†]See Chapter 4 for clarification on dependent and independent variables.
[‡]A different result is obtained if gas expands irreversibly.

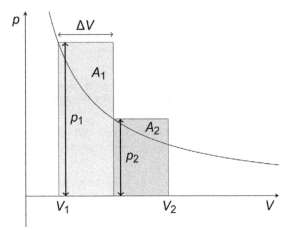

Figure 8.7 We can get a better approximation of the area between V_1 and V_2 by splitting the area in two, with area $A = A_1 + A_2 = p_1 \Delta V + p_2 \Delta V$, but there is still an appreciable error in the end result.

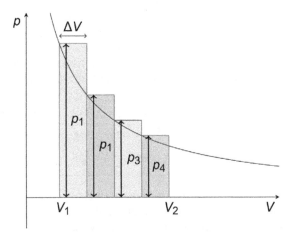

Figure 8.8 We can refine our approximation still further with more rectangles of constant width ΔV. The area becomes $A = A_1 + A_2 + A_3 + A_4 = (p_1 + p_2 + p_3 + p_4)\Delta V$. Again, while better, it is still not perfect.

We can see from Figure 8.5 to 8.8 that as we increase the number of rectangles, we get a better approximation to our area enclosed. We can express a formula for the area, as:

$$
\begin{aligned}
\text{Area}, A &= A_1 + A_2 + A_3 + A_4 \\
&= p_1 \Delta V + p_2 \Delta V + p_3 \Delta V + p_4 \Delta V \\
&= \left(p_1 + p_2 + p_3 + p_4 \right) \Delta V \\
&= \sum_{i=1}^{4} p_i \Delta V
\end{aligned}
\tag{8.14}
$$

We have introduced a new mathematical symbol, Σ. This is a way to shorten a summation series, as we see in eqn (8.14). The explanation of this is given in Key Concept 8.4.

---

❶ Key Concept 8.4 Mathematical Summations

There are many series in mathematics, and we need a simple way to illustrate these. One such series is a *summation*. Instead of writing a summation series as:

$$S_{total} = s_1 + s_2 + s_3 + s_4 + s_5 + s_6 + s_7$$

or even:

$$S_{total} = s_1 + s_2 + \cdots + s_7$$

we can notate it as:

$$S_{total} = \sum_{i=1}^{7} s_i$$

Translating this, our notation is saying, 'Add up the values of s_i from the first value of $i = 1$ to the last value of $i = 7$'. The value below the summation sign (the Greek upper-case letter sigma) Σ gives the lower limit of the summation, while the value above the summation gives the upper limit of the summation.

In this way, we can quickly notate a summation (or a subtraction by putting a negative value in, as $\sum_{i=1}^{7}(-r_i)$) for a large number of values, while using comparatively little space on a page.

We notice that the more rectangles we use, and hence the smaller ΔV becomes, the closer our calculated area will come to the actual area enclosed between these two values. Indeed, it should become apparent that our calculated area will become *exactly* the area enclosed *in the limit* as $\Delta V \rightarrow 0$ (see Key Concept 7.1). This principle is shown in Figure 8.9; the overall sum of the individual areas approaches the area actually enclosed by the curve as the slice width decreases. Remember that we have seen this principle of reducing step sizes before when considering differentiation (Chapter 7).

The challenge is to determine a formula for the area between two limits. If we continue the pressure–volume (pV) example, we would typically want to determine the work done as an ideal gas expands from one volume (V_1) to a larger volume (V_2). Remembering that the units of pV are congruent with energy, multiplying the pressure by the volume (*i.e.* finding the area enclosed by the curve) will give

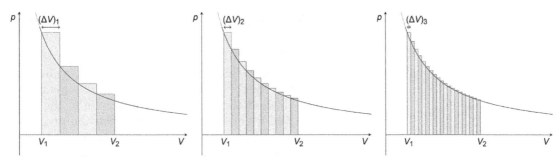

Figure 8.9   As we reduce the thickness of our 'slices' (ΔV), the closer the total area enclosed comes to the actual value for the area under the curve.

us an energy. There are a couple of slight problems with this consideration, however.

- Chemically, it is not reasonable to quantify the energy of a system solely from the product of a single pressure and volume; this value would have no physical significance.

- Mathematically, since $p \propto 1/V$, the curve never crosses either axis (remember that we are considering an ideal gas); it is not possible to evaluate the total area enclosed by the curve at a single value of V. This would represent a rectangular area enclosed between the prescribed point of interest and the origin of the graph, again having no physical significance.

Because of these problems, to find the work done as the gas expands, we need to *find the area between limits*; *i.e.* to find the area enclosed by the curve between two limits. We will refer to these limits as V_1 and V_2. The chemical question then becomes, 'What is the work done as a gas expands from V_1 to V_2?'

If we slice up the area as shown in Figure 8.9, we will gain a *series* of rectangles, where each rectangle represents 'a little bit of work done'. Adding up all of these pieces will give us an approximation to the total area enclosed in that area. To use our summation notation (Key Concept 8.4), we could write our expansion question as:

$$\text{Work done} = \sum_{i=V_1}^{V_2} \Delta w_i = -\sum_{i=V_1}^{V_2} p_i \Delta V \qquad (8.15)$$

Remember that we want to *translate* this to understand what it is saying. It is saying, 'The change in energy (work, Δw) is equal to the sum of all the individual "little bits of work done" between V_1 and V_2.' Notice that a negative sign has appeared; we consider all changes with respect to the system, and an increase in volume ($+\Delta V$) represents work done *on the surroundings*, therefore a *loss of energy* from the system.

What eqn (8.15) does not tell us is the *step size* of ΔV; the smaller the step size, the more terms appear in this summation, but the more accurate the result becomes in the context of determining the area enclosed (Figure 8.9). It also asks us to vary p_i (the 'i'th pressure value) within limits defined in terms of the volume, V. We therefore need to make a substitution using the ideal gas law:

$$p_i = \frac{nRT}{V_i} \tag{8.16}$$

We then substitute eqn (8.16) into eqn (8.15) to obtain the expression for the work done under reversible conditions:

$$\text{Work done} = -\sum_{i=V_1}^{V_2} \frac{nRT}{V_i}\Delta V = -nRT\sum_{i=V_1}^{V_2} \frac{1}{V_i}\Delta V \tag{8.17}$$

We have already said that this sum becomes equal to the area enclosed by the curve *in the limit* as $\Delta V \to 0$. This necessitates a change in notation; our summation will change (as we now have a sum with infinitely many steps) and we are now using the infinitesimal dV as our 'infinite summation' becomes an 'integral'

$$\text{Work done} = -\int_{V_1}^{V_2} p\,dV = -nRT\int_{V_1}^{V_2} \frac{1}{V}dV \tag{8.18}$$

This expression looks very similar to the general integration expression introduced in Key Concept 8.1, and this is exactly the process that we have uncovered. However, we have introduced a new aspect to the notation, namely the idea of *limits*. The result of this is something we know as a *definite integral*.

8.4.2   *The Definite Integral*

We saw in Key Concept 8.2 that, through the course of integration, we will obtain a *constant of integration* which can only be evaluated if we have more information about our curves (Example 8.1). By using a definite integral, we introduce a change to the notation.

For a general quantity Q, which varies with another property, x, the area enclosed between two points a and b may be visualised as shown in Figure 8.10.

Finding this directly is tricky; remember that we had to deal with a constant of integration, which brings elements of uncertainty.

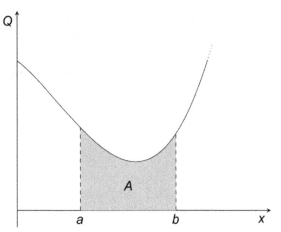

Figure 8.10   An area enclosed between a curve and the x-axis between the limits *a* and *b*.

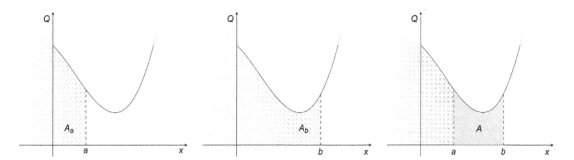

Figure 8.11   A definite area *A* can be found by finding the difference in the 'area up to *a*', A_a, and the 'area up to *b*', A_b. We don't know where to start measuring A_a or A_b (this information is wrapped up in the constant of integration), but they will both be subject to the *same* constant.

However, we know that the 'area up to limit *a*' and the 'area up to limit *b*' *will both have the same constant of integration.*

We also know that to find the area *A*, we need to find a *difference* in areas; *i.e.* the difference between the area up to *b* and the area up to *a* (Figure 8.11).

We can see that the area we want is equal to the difference between the areas A_a and A_b, and we write this as

$$A = A_b - A_a$$

$$A = \left[\int Q \mathrm{d}x \right]_b - \left[\int Q \mathrm{d}x \right]_a \qquad (8.19)$$

$$\equiv \int_a^b Q \mathrm{d}x$$

The full application of the definite integral is explained further in Key Concept 8.5 and demonstrated in Example 8.2.

---

❶ Key Concept 8.5 The Definite Integral

Suppose a quantity Q is defined as:

$$Q = 3x^2 + 2$$

To find the area enclosed between $x = 1$ and $x = 3$, we set up our definite integral as:

$$\int_1^3 Q\,dx = \int_1^3 \left(3x^2 + 2\right)dx$$

Translating this, we are 'integrating Q *from* $x = 1$ *to* $x = 3$ with respect to x'. As this is an addition, we can integrate each phrase on its own:

$$\int 3x^2 dx = \frac{3x^{(2+1)}}{(2+1)} = x^3 + C, \qquad \int 2dx = \frac{2x^{(0+1)}}{0+1} = 2x + C$$

We can combine these together (and the constant of integration) as:

$$\int_1^3 \left(3x^2 + 2\right)dx = \left[x^3 + 2x + D\right]_1^3$$

Note that we have combined the two constants of integration into a different constant D, and that we have carried forward the limits 1 and 3 on the brackets. Remember that the definite integral is a *difference* between two areas, and that in this notation we are integrating *from* 1 *to* 3. To evaluate this, therefore, we substitute our values for $x = 1$ and $x = 3$ into the integral, as:

$$\left[x^3 + 2x + D\right]_1^3 = \left[x^3 + 2x + D\right]_3 - \left[x^3 + 2x + D\right]_1$$
$$= \left[3^3 + 2 \times 3 + D\right] - \left[1^3 + 2 \times 1 + D\right]$$
$$= \left[27 + 6 + D\right] - \left[1 + 2 + D\right]$$
$$= (33 + D) - (3 + D)$$
$$= 33 + D - 3 - D$$
$$= 30$$

- The constant of integration is eliminated (it has the same value for both integrals, so when we find the difference it is reduced to zero).
- We *always* subtract the lower limit from the upper limit. If instead the integral in the example is $\int_3^1$, we should end up with a value of −30 instead of +30. In mathematical terms

$$\int_a^b f'(x)\,dx = \left[f(x)\right]_a^b \equiv f(b) - f(a)$$

Example 8.2 Calculating Work Done in a Reversible Expansion

Q: Calculate the work done when exactly one mole of an ideal gas expands reversibly from 1.00 dm³ to 3.00 dm³ at a constant temperature of 298 K.

A: As this is an ideal gas, we can use the ideal gas equation $pV = nRT$. From the discussions in Section 8.5 and elsewhere in the book, we have seen that the area under a p–V curve will have units equivalent to energy, giving the work done for any change in volume or pressure.

To find this area, we need to determine a definite integral between our limits of 1.00 dm³ and 3.00 dm³.

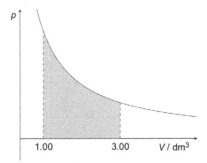

The integral we need to carry out, therefore, is:

$$w = -\int_{V_i}^{V_f} p\,dV$$

where we are integrating from our initial volume, V_i (in this case, 1.00 dm³), to our final volume, V_f (in this case, 3.00 dm³). Remember that the negative sign comes in as an *increase* in volume represents a *decrease* in total energy of the system; the system *has done work* on the surroundings. As discussed, we cannot integrate p directly, so we rearrange the gas equation and substitute for p:

$$pV = nRT$$

$$p = \frac{nRT}{V}$$

We can now calculate the integral using this substitution:

$$w = -\int_{V_i}^{V_f} p\,dV \equiv -\int_{V_i}^{V_f} \frac{nRT}{V}\,dV \equiv -nRT\int_{V_i}^{V_f} \frac{1}{V}\,dV$$

We now apply the principles of definite integrals:

$$w = -nRT \int_{V_i}^{V_f} \frac{1}{V} dV$$

$$= -nRT \left[\ln V \right]_{V_i}^{V_f}$$

Finally, we can apply the rules of logarithms (Chapter 1) to obtain a general expression for this definite integral:

$$w = -nRT \left[\ln V \right]_{V_i}^{V_f}$$

$$= -nRT \left[\ln V_f - \ln V_i \right]$$

$$= -nRT \ln \left(\frac{V_f}{V_i} \right)$$

In the context of our calculation, we know that our initial volume, V_i, is 1.00 dm³ and our final volume, V_f, is 3.00 dm³, so we can substitute these values into our general expression:

$$w = -nRT \ln \left(\frac{V_f}{V_i} \right)$$

$$= -1 \text{ mol} \times 8.314 \text{ J K}^{-1} \text{ mol}^{-1} \times 298 \text{ K} \times \ln \left(\frac{3.00 \text{ dm}^3}{1.00 \text{ dm}^3} \right)$$

$$= -2.72 \times 10^3 \text{ J}$$

This tells us that the system has lost energy or, to put another way, has *done work on the surroundings.*

Since the units of dm³ cancel here you don't need to undergo a unit conversion.

When a value is termed as *exact* it can be considered to have infinite significant figures.

? Exercise 8.4 Determining the Value of Definite Integrals

Evaluate the following definite integrals:

1. $w = \int_0^3 x^2 dx$

2. $b = \int_{-\frac{\pi}{2}}^{\frac{\pi}{2}} \cos(a) da$

3. $y = \int_0^2 e^{0.5z} + 2 \, dz$

4. $r = \int_1^5 \frac{2}{s} + 2s + 3 \, ds$

5. $w = \int_{0.24}^{0.96} \frac{1}{V} dV$

6. $x = \int_0^\pi \cos 2y + \sin y \, dy$

7. $f(g) = \int_0^2 g^{-1/2} + 2 \, dg$

8. $z = \int_0^5 1 + y + y^2 + y^3 \, dy$

Determine integrated forms of the equation for the following definite integrals:

9. $\ln p = \int_{T_1}^{T_2} \dfrac{\Delta_{vap}H}{RT^2}\,dT$

10. $w = \int_{V_1}^{V_2} \dfrac{nRT}{V}\,dV$

11. $\Delta H = \int_{T_1}^{T_2} C_p\,dT$

12. $\int_{H_{m,T_1}}^{H_{m,T_2}} dH_m = \int_{T_1}^{T_2}\left(a + bT + \dfrac{c}{T^2}\right)dT$

8.5 Integration in Context

A standard model in quantum chemistry is the 'particle in a one-dimensional box'. In this model, a particle is constrained to move in a single dimension and is only permitted to carry kinetic energy (*i.e.* its potential energy is zero). Using this crude model, we can make reasonable predictions of the electronic absorptions of conjugated hydrocarbons (*e.g.* butadiene, 1,3,5-hexatriene); however, in constructing this model, we start with a *wave function* and must *normalise* this against known probability factors.

The mathematical principles of the model are simple; for a box of length L:

- The wave function must be continuous and single-valued.
- The wave function must be zero at the edges of the box.
- The particle must exist *somewhere* in the box.

For the sake of this discussion, we can model a wave function as a sine wave inside the box. Examples of some wave functions fulfilling these conditions are shown in Figure 8.12.

These wave functions are described by the general form $A\sin(kx)$. The value of k can be easily quantified.

- At $x = L$, the value of the wave function is zero.
- Therefore, $A\sin(kL) = 0$.
- Sine functions are equal to zero when acting on values 0, π, 2π, ..., $n\pi$.
- Therefore $kL = n\pi$.
- Therefore $k = {}^{n\pi}/_L$.

This gives us a mathematical description for our wave function:

$$\Psi = A\sin\left(\frac{n\pi x}{L}\right) \tag{8.20}$$

However, our pre-function scaling factor (or 'amplitude'), A, has yet to be determined. Without this, we don't have a wave function

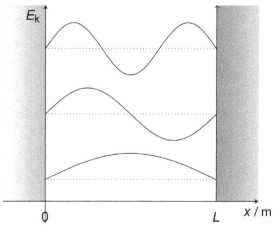

Figure 8.12   Example wave functions for the particle in a one-dimensional box. The sine functions shown are of the form $A \sin(kx)$, each is single-valued and the value of the wave function is zero at the edges of the box. (The wave function zero value is indicated by the dotted line.)

> 💬 In quantum chemistry, we work in radians, not degrees. See Section 5.1.1 for further discussion on radians.

we can rely on for making predictions in our chemical systems. To find a value of A, that is to find the correct scaling factor for our wave function, we need to carry out a process called *normalisation*.

8.5.1   Normalising a Wave Function by Integration

When we normalise a function, we are comparing it with a known value. In the case of the wave function, we *know* that the particle must be in the box (otherwise our model is useless); therefore, the *probability* of finding the particle must be equal to 1.

How does this translate to our wave function? We know that the probability of finding a particle is proportional to the square of the wave function (Ψ^2),[§] and the area under the curve of Ψ^2 $(\Psi^*\Psi)$ plotted against position will be equal to 1. The integral we are therefore trying to find already has a solution (eqn (8.21)); however, the precise nature of the constants in the integral are unknown.

$$\int_0^L \Psi^*\Psi \, dx = 1 \tag{8.21}$$

If we 'translate' eqn (8.21), we are being told that 'The integral of the probability (wave function squared) from position $x = 0$ to position $x = L$ (*i.e.* the whole of the inside of the box) with respect

---

[§]Formally, as wave functions are complex in nature, this will be the product of the wave function and its complex conjugate, $\Psi^*\Psi$. See Chapter 9 for further information on complex numbers and complex conjugates.

to position is equal to one.' We already have an expression for the wave function, so let us now put this into the integral:

$$\int_0^L \left[A\sin\left(\frac{n\pi x}{L}\right)\right]^2 dx = 1$$

$$\int_0^L A^2 \sin^2\left(\frac{n\pi x}{L}\right) dx = 1 \qquad (8.22)$$

$$A^2 \int_0^L \sin^2\left(\frac{n\pi x}{L}\right) dx = 1$$

Now we need to evaluate our definite integral and rearrange the equation to find a value for A. However, integrating $\sin^2(kx)$ is not trivial, as this is a nested function. In this case, we need to use a *substitution*. There are a number of substitutions we can use for trigonometric functions to simplify this as there are a wealth of trigonometric identities (see Chapter 5). In this case, we use the trigonometric identity for a double-angle formula; we need a formula which includes the $\sin^2(Bx)$ function but no other 'tricky' functions. We can see that the $\cos(2\theta)$ formulation will work as a substitution, since it includes a $\sin^2(kx)$ component:

$$\cos(2\theta) = \cos^2\theta - \sin^2\theta$$
$$= 2\cos^2\theta - 1$$
$$= 1 - 2\sin^2\theta \qquad (8.23)$$

Recall:

$\sin^2 x + \cos^2 x = 1$

We see from eqn (8.23) that there are three identities for $\cos(2\theta)$, two of which have a $\sin^2\theta$ relation. The first of these also includes a $\cos^2\theta$ function, so this will not simplify our integration. The third of these, however, gives us an expression which we can use. We can rearrange this to generate the substitution we need for $\sin^2\theta$ in our integration:

$$\cos(2\theta) = 1 - 2\sin^2\theta$$

$$\sin^2\theta = \frac{1 - \cos(2\theta)}{2} \qquad (8.24)$$

When we compare eqn (8.24) with eqn (8.22), we replace the term θ with $n\pi x/L$. After this substitution, our integral becomes:

$$A^2 \int_0^L \sin^2\left(\frac{n\pi x}{L}\right) dx \equiv A^2 \int_0^L \frac{1 - \cos\left(\frac{2n\pi x}{L}\right)}{2} dx \equiv \frac{A^2}{2} \int_0^L 1 - \cos\left(\frac{2n\pi x}{L}\right) dx$$

$$(8.25)$$

Remembering that $\cos(Bx)$ integrates to $\left(\sin(Bx)\right)/B$ we now carry out this integration:

$$\frac{A^2}{2}\int_0^L 1-\cos\frac{2n\pi x}{L}dx = 1$$

$$\frac{A^2}{2}\left[x-\frac{\sin\left(\dfrac{2n\pi x}{L}\right)}{\left(\dfrac{2n\pi}{L}\right)}\right]_0^L = 1$$

$$\frac{A^2}{2}\left[x-\frac{L}{2n\pi}\times\sin\frac{2n\pi x}{L}\right]_0^L = 1 \tag{8.26}$$

This seems to be a complicated result, however, consider the boundary conditions; when $x = 0$, we calculate $\sin(0)$, which is equal to zero. When $x = L$, we need to calculate $\sin(2n\pi)$; as this is cyclic, and n is an integer, $\sin(2\pi) = 0$, $\sin(4\pi) = 0$, *etc.* Therefore, the $\sin\left(2n\pi x/L\right)$ component will always reduce to zero, and our definite integral evaluates as follows:

$$\frac{A^2}{2}\left[x-\frac{L}{2n\pi}\times\sin\frac{2n\pi x}{L}\right]_0^L = 1$$

$$\frac{A^2}{2}\times\left(\left[L-\frac{L}{2n\pi}\times\sin\frac{2n\pi L}{L}\right]-\left[0-\frac{L}{2n\pi}\times\sin\frac{2n\pi\times 0}{L}\right]\right) = 1$$

$$\frac{A^2}{2}\times(L-0) = 1$$

$$\frac{A^2}{2}\times L = 1$$

$$A = \sqrt{\frac{2}{L}} \tag{8.27}$$

Remember the sine of any integer multiple of 2π evaluates to zero, matching our boundary condition.

This gives us the final wave function for the particle in a one-dimensional box, as:

$$\Psi_x = \sqrt{\frac{2}{L}}\sin\frac{2n\pi x}{L} \tag{8.28}$$

This is one way of using a substitution to simplify an integration; the next method we show is the formal integration by substitution process.

8.5.2 Integration by Substitution

The next example of integrating by substitution is the formal 'integration by substitution' process. It can be thought of as the reverse process to the 'chain rule' (see Chapter 7) and applies if we can write the integral as a nested function. The simplest way to visualise this is to look at a simple example. Let's suppose that the force, F, in a complex field can be related to a separation, r by the relationship:

$$F = (3r+6)^4 \qquad (8.29)$$

We know that the potential, V, is the integral of the force with respect to the separation r, but it is not immediately clear how we integrate this as the integration is not trivial. However, we have already said that we are looking for an integral complement for differentiation using the 'chain rule'. Therefore, we can specify a new function:

$$V = \int F dr \qquad (8.30)$$

$$= \int (3r+6)^4 \, dr$$

$$= \int u^4 dr \text{ where } u = (3r+6)$$

While this now looks much simpler, we cannot directly integrate a variable u with respect to a different variable r; we need to identify a way to integrate with respect to u and to replace the dr with a du. Let's firstly find the first derivative of u with respect to r:

$$u = 3r+6$$

$$\frac{du}{dr} = 3 \qquad (8.31)$$

We can now 'separate the variables' (see Section 8.3.1) and this immediately shows us a relationship between dr and du:

$$\frac{du}{dr} = 3$$

$$dr = \frac{1}{3}du \qquad (8.32)$$

We can replace dr in our integral with this ⅓ du, and we can now complete our integration:

$$V = \int u^4 dr \text{ where } u$$

$$= \int u^4 \frac{1}{3}du$$

$$= \frac{1}{3}\int u^4 du$$

$$= \frac{1}{3} \times \frac{u^5}{5} + C$$

$$= \frac{u^5}{15} + C$$

$$= \frac{(3r+6)^5}{15} + C \qquad (8.33)$$

Note that again we must consider that we have a constant of integration. As an alternative approach, we could decide to expand the brackets and integrate each term individually, but this becomes a huge amount of work.

This approach to integration by substitution can be made if we can identify a chained function for which the substituted function u differentiates to a constant value. This means that we can make a straightforward substitution for u and dr. This technique relies on the same principle as the chain rule, namely that:

$$du = \frac{du}{dr} \times dr \qquad (8.34)$$

where it can be seen how the 'chain of cancellation' occurs.

❓Exercise 8.5 Integration by Substitution

Using the substitution shown, integrate the following:

1. $\int \sin(g)\cos(g)dg$; let $u = \sin(g)$

2. $\int 2\sin^4(x)\cos(x)dx$; let $u = \sin(x)$

3. $\int ae^{5a^2}da$; let $u = 5a^2$

4. $\int (z^2+1)(z^3+3z+1)^3 dz$; let $u = (z^3+3z+1)$

5. $\int \frac{x}{(x+1)^{1/2}}dx$; let $u^2 = x+1$

6. $\int (16b^3 + 12b^2 + 8b + 4\cos(b))(b^4 + b^3 + b^2 + 1 + \sin(b))^3 db$;

   let $u = (b^4 + b^3 + b^2 + 1 + \sin(b))$

7. $\int_0^\pi \sin(\theta + a)d\theta$; let $u = \theta + a$

8. $\int_a^b \frac{6x}{(3x^2+2)}dx$; let $u = 3x^2 + 2$

For Question 7, as you substitute u in here, you can change your limits and save yourself from back substituting later.

By determining a suitable substitution, integrate the following:

9. $\int 2x \cos(x^2)dx$

10. $\int_0^{\frac{\pi}{2}} \cos(a+\pi)e^{\sin(a+\pi)} da$

11. $\int \frac{(\cos(b))}{(1+\sin(b))}db$

12. $\int s^2 e^{s^3} ds$

13. $\int \frac{(2\ln(z))}{z}dz$

14. $\int \frac{x}{(\sqrt{(3x^2+2)^3})}dx$

8.5.3 Integration by Parts

The last example of more complex integration involves a process called 'integration by parts'. Whereas 'integration by substitution' was the reverse process of 'differentiation by the chain rule', 'integration by parts' can be seen as the reverse process for 'differentiation by the product rule'. Remember that the product rule is used for compound functions where u and v are both functions of our variable (in this case, we will use the variable x); for example, u might be equal to $\sin(kx)$, while v might be equal to x^2, with our overall function of x being $x^2 \sin(kx)$. To differentiate this using the product rule, we applied the relationship:

$$\frac{d(uv)}{dx} = u\frac{d(v)}{dx} + v\frac{d(u)}{dx} \tag{8.35}$$

We used this rule in Chapter 7 to differentiate appropriate compound functions; rearranging this produces:

$$u\frac{d(v)}{dx} = \frac{d(uv)}{dx} - v\frac{d(u)}{dx} \tag{8.36}$$

While this looks like a somewhat arbitrary rearrangement, when we integrate both sides of this expression, we get an altogether more useful result:

$$\int u\frac{d(v)}{dx}dx = uv - \int v\frac{d(u)}{dx}dx \tag{8.37}$$

Put simply, if we can identify the two component functions (u and $d(v)/dx$) of our compound function, we can integrate one to get v and differentiate the other to get $d(u)/dx$; putting the two parts back together, as shown in eqn (8.37), we can identify an alternate expression for the integral. This process is shown in Example 8.3.

Example 8.3 Integrating by Parts in Practice

Consider the following compound function:

$$t = 2\theta \times \cos(3\theta)$$

If we want to integrate this expression, we need to integrate by parts, as the two functions are multiplied together. We need to identify the two parts; one of which is in its integrated form, while the other is in its derivative form:

$$u = 2\theta \qquad \frac{dv}{d\theta} = \cos(3\theta)$$

When we put this back together, as shown in eqn (8.37), we need to find the derivative of the u part, while we need to identify the integral of the $dv/d\theta$ part:

$$u = 2\theta \qquad \frac{dv}{d\theta} = \cos(3\theta)$$

$$\frac{du}{d\theta} = 2 \qquad v = \int \cos(3\theta)d\theta$$

$$= \frac{1}{3}\sin(3\theta) + C$$

When deciding which term is u and which v you should be looking to identify a term which, when differentiated (once or sometimes more than once), becomes simpler. So here 2θ differentiates to be 2 and the resulting term is now easier to integrate.

Now we put the parts back together:

$$\int u \frac{d(v)}{d\theta} d\theta = uv - \int v \frac{d(u)}{d\theta} d\theta$$

$$= \left[2\theta \times \left(\frac{1}{3} \sin(3\theta) + C \right) \right] - \int \left[\left(\frac{1}{3} \sin(3\theta) + C \right) \times 2 \right] d\theta$$

$$= \left[\frac{2\theta \sin(3\theta)}{3} + 2\theta C \right] - \int \left[\frac{2 \sin(3\theta)}{3} + 2C \, d\theta \right]$$

$$= \left[\frac{2\theta \sin(3\theta)}{3} + 2\theta C \right] - \left[\frac{2 \cos(3\theta)}{9} + 2\theta C + D \right]$$

$$= \frac{2\theta \sin(3\theta)}{3} + \frac{2 \cos(3\theta)}{9} - D$$

The core of 'integration by parts' is identifying the two 'parts' of the compound function; one of them we integrate, the other we differentiate. The key to this process, however, is that the 'part' which we differentiate *ideally* should differentiate to a constant value; this is the only way that, when we put our parts back together as shown in eqn (8.37), the second integral on the right-hand side becomes simpler. If we were to look at Example 8.3 again, however instead we swap the 'parts' around, and integrated and differentiated the opposite parts, we would find that our integral would not simplify and we would be left with a more complex integral to solve. Fortunately, this becomes apparent quickly and before we get too far with our integration.

As a general rule of thumb, the 'part' which is differentiated should be a simple polynomial of the variable (*i.e.* for variable r, the 'part' to differentiate should be in the form r^n, where n is an integer value), as this will (eventually) simplify to a constant value. Any exponential component, trigonometric component, *etc.* should be the 'part' which you integrate, as this will not become simpler with differentiation.

8.5.4   Standard Integrals

You may find that with some integrals you need to use a range of techniques shown to undertake the integration, and may even need to repeat the 'integration by parts' process until the integral becomes simple enough to complete. However, this can become arduous and there are a great many functions which already have an integral available for use. These are called 'standard integrals' and serve to limit the amount of 'reinvention' required. They may not be in the immediate form you need however, so it is worth checking that you are able to use these. Example 8.4 shows how to adapt one type of standard integral for use; similar approaches can be used for other standard integrals.

Example 8.4 Using a Standard Integral

Q: Integrate the function:

$$w = 3r^2 e^{5r}$$

A: To integrate this expression, we would need to set up an integral by parts, and then to solve the new integral we would need to integrate by parts a second time. This can be exceptionally time-consuming, particularly for very high powers. Fortunately, a standard integral exists for this expression, and takes the form:

$$\int x^2 e^{ax}\, dx = \left(\frac{x^2}{a} - \frac{2x}{a^2} + \frac{2}{a^3}\right)e^{ax} + C$$

To use this standard integral we need to recognise a number of things in relation to our function:

- The variable in our function is r, not x.
- We have a scaling factor of 3 in our function relative to this standard integral.
- The index of the exponent has a value of 5 in place of a.

Recognising these factors, we can now apply the standard integral:

$$\int 3r^2 e^{5r}\, dr \equiv 3\int r^2 e^{5r}\, dr$$

$$3 = \left(\frac{r^2}{5} - \frac{2r}{5^2} + \frac{2}{5^3}\right)e^{5r} + C$$

? Exercise 8.6 Integration by Parts

Using the listed values of the parts u and v, determine the integrals of the following:

1. $\int x\sin(2x)\,dx; u = x$ and $dv = \sin(2x)\,dx$

2. $\int 3ae^{7a}\,da; u = 3a$ and $dv = e^{7a}\,da$

3. $\int \ln(\rho)\,d\rho; u = \ln\rho$ and $dv = 1\,d\rho$

4. $\int ze^{-z}\,dz; u = z$ and $dv = e^{-z}\,dz$

5. $\int a\ln a\,da; u = \ln a$ and $dv = a\,da$

6. $\int x^4 \ln x\,dx; u = \ln x$ and $dv = x^4\,dx$

By determining suitable parts, determine the following integrals:

7. $\int 4b\cos(b)\,db$

8. $\int x^2 e^{2x}\,dx$

9. $\int (g+1)\sin(g)\,dg$

10. $\int 3\ln(\phi)\,d\phi$

11. $\int \gamma\cos(\pi\gamma)\,d\gamma$

12. $\int_0^\pi \theta\sin(2\theta)\,d\theta$

8.6 Summary

We initially introduced the concept of integration as the inverse function for differentiation, a way to reverse the processes we encountered in Chapter 7. However, we have additionally shown that the application of integration allows the determination of areas enclosed in graphs. In the context of chemistry, this is of paramount importance, particularly when we have quantities which are interrelated but do not have a linear relationship (*e.g.* pV relationships) or we need to normalise a function, u, such as a wave function or the Maxwell–Boltzmann distribution.

Summary: Chapter 8

An 'antiderivative' is the process of finding the curve of a function when we know the rate of change of the variables at a given point. Integration is the process of using the 'antiderivative' to identify the area enclosed by the curve exactly, and in using a definite integral we avoid the 'constant of integration' problem.

Table of Some Standard Integrals

$f'(x) \equiv y$	$f(x) = \int y\, dx$	Condition
A	$x + C$	
Ax^n	$\dfrac{A}{n+1}x^{n+1} + C$	$x \neq -1$
$\dfrac{A}{x} \equiv Ax^{-1}$	$A\ln x + C$	
$\dfrac{A}{Bx+d}$	$\dfrac{A}{B}\ln(Bx+d) + C$	
e^{Ax}	$\dfrac{1}{A}e^{Ax} + C$	
A^x	$\dfrac{A^x}{\ln A} + C$	
$A\ln x$	$A(x\ln x - x) + C$	
$\sin Ax$	$-\dfrac{1}{A}\cos Ax + C$	
$\cos Ax$	$\dfrac{1}{A}\sin Ax + C$	
$\tan Ax$	$-\dfrac{1}{A}\ln(\cos Ax) + C \equiv \dfrac{1}{A}\ln(\sec Ax) + C$	$-\dfrac{\pi}{2} < x < \dfrac{\pi}{2}$

$f'(x) \equiv y$	$f(x) \equiv \int y\, dx$	Condition
$\dfrac{A}{x^2 + B^2}$	$\dfrac{A}{B}\tan^{-1}\left(\dfrac{x}{B}\right) + C$	$B > 0$
$Ax^2 e^{Bx}$	$A\left(\dfrac{x^2}{B} - \dfrac{2x}{B^2} + \dfrac{2}{B^3}\right)e^{Bx} + C$	

Note: This list is not exhaustive; many other standard integrals are available from a wide range of sources available both online and in print.

Integration by Parts and by Substitution

In the following examples, the terms u and v indicate different functions of x.

	$f'(x) \equiv y$	$f(x) \equiv \int y\, dx$
Integration by substitution	$y = f(u);\ ug(x)$	$\displaystyle\int u\frac{du}{dx}dx = \int u\, du$
Integration by parts	$y = u\dfrac{dv}{dx}$	$\displaystyle uv - \int v\frac{du}{dx}dx$

❓Exercise 8.7 Integration Problems in a Chemical Context

Each of the following questions is an example of integration in context in chemistry. Try to solve the problems mathematically. If you haven't met the topic yet in your lectures, don't worry; in this case, it is the maths which matters!

1. A student theorises that lithium can be reacted to form dilithium in a third-order reaction, such that:

$$\frac{d[\text{Li}]}{dt} = -k[\text{Li}]^3 \tag{8.38}$$

   (a) By using appropriate limits, determine the integrated form of the rate equation for this reaction.
   (b) Sketch a suitable straight-line ($y = mx + c$) graph which would give a linear relationship for determining the rate constant k.
   (c) Determine the units of the rate constant k.
   (d) Determine an equation for the half-life of lithium in this reaction.

2. The work done when a gas expands reversibly is given by:

$$\frac{\delta w}{\delta V} = -\frac{nRT}{V} \tag{8.39}$$

By use of suitable limits, derive an expression for the expansion work.

The expression can include $\ln p$.

3. For phase boundaries, the Clapeyron equation says how the pressure and temperature for a phase boundary are linked:

$$\frac{\mathrm{d}\ln p}{\mathrm{d}T} = \frac{\Delta_{vap}H}{RT^2} \tag{8.40}$$

Determine an expression for how the pressure varies with temperature.

Sketch the graph and don't try to cancel or multiply out any of the 'T's!

4. At very low temperatures, the absolute molar entropy of a substance is the area under the curve of the function $C_p/T = aT^3$. Determine the absolute molar entropy of AgO at 10 K, given that $a = 3.53 \times 10^{-3}$ J K^{-5} mol^{-1}.

5. The differential form of the Arrhenius equation is:

$$\frac{\mathrm{d}\ln k}{\mathrm{d}T} = \frac{E_a}{RT^2} \tag{8.41}$$

(a)   Integrate this expression in order to determine the integrated form of this equation, and arrange it so that k is the subject.

(b)   Sketch an appropriate graph to determine the activation energy, E_a.

6. The force of attraction, F, between two charged particles is given by Coulomb's law:

$$F = \frac{q_1 q_2}{4\pi\varepsilon_0 r^2} \tag{8.42}$$

where q_1 and q_2 are the charges on particles 1 and 2 respectively, r is the separation between the two particles and all other symbols are constants.

Derive an expression for the potential energy of the particles, given that it is the integral of the force with respect to distance, r.

Solutions to Exercises

Solutions: Exercise 8.1

1. $f(x) = \dfrac{x^3}{3} + \dfrac{x^2}{2} + 7x + C$

2. $f(p) = \dfrac{1}{p^3} + C$

3. $a = -\dfrac{b^{-2}}{2} + 2b^{1/2} + C$

4. $r = \dfrac{3}{5}s^{5/3} - 3s^{-1} + s^2 + as + C$

5. $n = \dfrac{\sin(m)}{2} + C$

6. $y = -4\sin\left(\dfrac{z}{4}\right) + 3z^2 + C$

7. $f(g) = -\dfrac{4}{a}e^{-ag} = -\dfrac{4}{ae^{ag}} + C$

8. $H(T) = aT + \dfrac{bT^2}{2} + \dfrac{cT^3}{3} + C$

9. $S = \dfrac{aT^3}{3} + C$

10. $V(r) = \dfrac{q_1 q_2}{4\pi\varepsilon_0 r} + C$

Solutions: Exercise 8.2

1. $y = \dfrac{2}{3}x^3 + C; \; C = 5$

2. $f(r) = -2\cos\left(\dfrac{r}{2}\right) + C; \; C = 2$

3. $p(q) = \dfrac{e^{3q}}{3} + \dfrac{q^2}{4} + C; \; C = \dfrac{2}{3}$

4. $a = \dfrac{3}{2}\sin(2b) + C; \; C = 3$

5. $z = \dfrac{w^4}{4} + \dfrac{2w^{3/2}}{3} + 2w^{1/2} + 2aw + C; \; C = 4a$

6. $m = -\cos\dfrac{(2n)}{2} + 2\sin\left(\dfrac{n}{2}\right) + \dfrac{e^{2n}}{2} + C; \; C = 0$

Solutions: Exercise 8.3

1. $y = 2\ln x + C$

2. $f(r) = \dfrac{r^3}{2} + \ln r + C; \; C = -\dfrac{1}{2}$

3. $p(q) = 2e^{q/2} + \dfrac{(\ln q)}{2} + C$

4. $z = \ln w + \dfrac{2w^{3/2}}{3} + 2w^{1/2} + C; \; C = \dfrac{7}{3}$

5. $b = \ln a + a\ln 7 + C$

6. $m = 4\ln x + 2\ln x + \dfrac{x^2}{2} + C$

Solutions: Exercise 8.4

1. $w = 9$

2. $b = 2$

3. $y = 2e + 2$

4. $r = 2\ln 5 + 36$

5. $w = \ln 4$

6. $x = -2$

7. $f(g) = 2\sqrt{2} + 4$

8. $z = 2585/12$

9. $\ln p = -\dfrac{\Delta_{vap}H}{R}\left(\dfrac{1}{T_2} - \dfrac{1}{T_1}\right)$

10. $w = nRT\ln\left(\dfrac{V_2}{V_1}\right)$

11. $\Delta H = C_p(T_2 - T_1)$

12. $\Delta H_m = \left(aT_2 + bT_2{}^2 - \dfrac{c}{T_2}\right)$
$\qquad - \left(aT_1 + bT_1{}^2 - \dfrac{c}{T_1}\right)$

Solutions: Exercise 8.5

1. $\sin^2\dfrac{(g)}{2} + C$

2. $\dfrac{2\sin^5(x)}{5} + C$

3. $\dfrac{e^{5a^2}}{10} + C$

4. $\dfrac{(z^3 + 3z + 1)^4}{12} + C$

5. $\dfrac{2(x+1)^{3/2}}{3} - 2(x+1)^{1/2} + C$

You might find it helpful to find $\dfrac{dx}{du}$, rather than $\dfrac{du}{dx}$.

6. $(b^4 + b^3 + b^2 + 1 + \sin(b))^4 + C$

7. Use the trigonometric identity for addition of angles (Key Concept 5.3). $2\cos(a)$

8. $\ln\left(\dfrac{(3b^2+2)}{(3a^2+2)}\right) + C$

9. Let $u = x^2$; $\sin(x^2) + C$

10. Let $u = \sin(a+\pi)$; $\dfrac{1-e}{e}$

11. Let $u = (1 + \sin b)$; $\ln(1 + \sin(b)) + C$

12. Let $u = s^3$; $\dfrac{e^{s^3}}{3} + C$

13. Let $u = \ln z$; $(\ln z)^2 + C$

14. Let $u = 3x^2 + 2$; $-\dfrac{-1}{3\sqrt{(3x^2+2)}} + C$

Solutions: Exercise 8.6

1. $\dfrac{\sin(2x)}{4} - \dfrac{x\cos(2x)}{2} + C$

2. $\dfrac{3e^{7a}}{7}\left(a - \dfrac{1}{7}\right) + C$

3. $\rho \ln \rho - \rho + C$

4. $-ze^{-z} - e^{-z} + C$

5. $\dfrac{a^2}{2}\ln a - \dfrac{a^2}{4} + C$

6. $\dfrac{x^5}{5}\ln x - \dfrac{x^5}{25} + C$

7. Let $u = 4b, dv = \cos b\,db$; $4b\sin(b) + 4\cos(b) + C$

8. Requires integrating by parts twice. Firstly, let $u = x^2, dv = e^{2x}dx$;
$$\int x^2 e^{2x}dx = \dfrac{x^2 e^{2x}}{2} - \int xe^{2x}dx$$
Then: let $u' = x, dv' = e^{2x}dx$;
$$= \dfrac{x^2 e^{2x}}{2} - \left[\dfrac{xe^{2x}}{2} - \int \dfrac{e^{2x}}{2}dx\right]$$
$$= \dfrac{e^{2x}}{2}\left(x^2 - x + \dfrac{1}{2}\right) + C$$

9. Let $u = (g+1)$, let $dv = \sin(g)dg$; $\sin(g) - (g+1)\cos(g) + C$

10. Let $u = \ln\phi$, let $dV = 3d\phi$; $3\phi(\ln\phi - 1) + C$

11. Let $u = \gamma$, let $dv = \cos(\pi\gamma)d\gamma$; $\dfrac{1}{\pi}\left(\gamma\sin(\pi\gamma) + \dfrac{\cos\pi\gamma}{\pi}\right) + C$

12. Let $u = \theta$, let $dv = \sin(2\theta)d\theta$; $-\dfrac{\pi}{2}$

Solutions: Exercise 8.7

1. (a) Let $[Li] \equiv [A]$; at time $t = 0$ s, $[A] = [A]_0$

$$\frac{1}{2[A]^2} = \frac{1}{2[A]_0^2} + kt$$

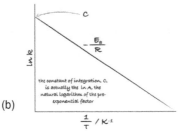

(b)

(c) Units of $k = \text{mol}^{-2} \text{ dm}^6 \text{ s}^{-1}$

(d) $t_{1/2} = \dfrac{3}{2[A]_0^2 k}$

2. Appropriate limits: At initial volume V_i, work is zero; at final volume V_f the work done is equal to w.

$$w = -nRT \ln\left(\frac{V_f}{V_i}\right)$$

3. $\ln p = -\dfrac{\Delta_{vap}H}{R}\dfrac{1}{T} + C$

4. $S = 8.82$ J K^{-1} mol^{-1}

5. (a) $k = e^{(-E_a/RT + C)}$

the constant of integration, C, is actually the ln A, the natural logarithm of the pre-exponential factor

(b)

6. Remember work done is in bringing particle *from* ∞ *to* r; Use these as limits in the integration to get a negative value. At $r = \infty$, $V(r) = 0$

$$V(r) = \int_{\infty}^{r} F dr$$

$$= \left[-\frac{q_1 q_2}{4\pi\varepsilon_0 r}\right]_{\infty}^{r} = -\frac{q_1 q_2}{4\pi\varepsilon_0 r}$$

Again, a sketch only needs to indicate the shape of the graph and the key characteristics you would expect to see.

Learning Points: What We'll Cover

- [ ] Complex numbers arising from the need to solve quadratic equations
- [ ] Arithmetic of complex numbers, including the complex conjugate
- [ ] Argand diagrams: graphical representation of a complex number
- [ ] Representing a complex number as an exponential and applying De Moivre's theorem to relate this to Argand diagrams
- [ ] Application of complex numbers in quantum chemistry and in the Schrödinger equation

Why This Chapter Is Important

- Quantum chemistry uses the 'wave–particle duality' when modelling the behaviour of electrons; complex numbers describe this wave behaviour.
- Complex numbers are an intrinsic characteristic of the wave function needed to solve the Schrödinger equation in quantum chemistry.
- Solutions to the Schrödinger equation can predict the shapes of atomic orbitals.

Complex Numbers: Quantum Mechanics and Solving the Schrödinger Equation

The idea of a 'complex number' sounds... *complex*; however, using many of the mathematical concepts with which you are probably already familiar, the treatment of a 'complex' number becomes straightforward. In this context, we use the term 'complex number' to mean 'a number with more than one part'. You will probably already be familiar with this use of the word when describing 'metal complexes' in transition metal chemistry.

This chapter forms a capstone chapter for this book; in our discussion of complex numbers, we will be using principles we covered when exploring vectors (Chapter 6) and trigonometry (Chapter 5), and when considering exponential functions (Chapter 1). Additionally, we will be showing how a complex number can be treated in the same way as any other number, subject to one simple additional consideration.

Despite their name, 'complex' numbers are not difficult to understand, and their use makes solving a range of chemical problems considerably more simple.

9.1 Complex Numbers in Chemistry

For the undergraduate chemist, complex numbers first appear in the area of quantum and computational chemistry, and students who focus on the boundary between physical chemistry and chemical physics will see their increasing prevalence in the models used to explore chemistry at this interface. Their appearance is due to the consideration of the 'wave–particle' nature of matter. In describing the behaviour of particles using wave mechanics, we end up with two separate—but inextricably linked—values. A 'complex number', with its two-component nature, allows us to consider these properties mathematically. First of all, let's introduce one of the most fundamental equations in quantum mechanics: the Schrödinger equation.

9.1.1 The Schrödinger Equation

The Schrödinger equation is deceptively simple in its form:

$$\hat{\mathbf{H}}\Psi = E\Psi \tag{9.1}$$

To understand its meaning, let's unpack what each term in eqn (9.1) represents:

$\hat{\mathbf{H}}$ The Hamiltonian *operator*, the mathematical instruction to perform on Ψ.

Ψ The wave function of the particle we are considering.

E The *eigenvalue* giving the magnitude of the quantity we are looking for.

To put this in plain English, the wave function Ψ is a full description of the behaviour of a particle or system; if we want to obtain information from it, we need to perform an *operation* on it. If we wish to determine the momentum of a particle, we use the Hamiltonian operator for momentum. This is simply a mathematical instruction to perform on the wave function. The property of the operator however is that it must always return the original wave function (hence the presence of Ψ on both sides of the equation) multiplied by a constant value (the eigenvalue, E). If we use the momentum operator, then the eigenvalue E will tell us the momentum of the particle; if we use the energy operator, then the eigenvalue E will give us the energy of the particle.

To fully compute an outcome of the Schrödinger equation, we need to better understand the form of a wave function, and to do this we need to be able to manipulate complex numbers.

9.1.2 The Need for Complex Numbers

One of the central concepts in quantum mechanics is the 'wave–particle duality'; the concept that bodies traditionally thought of as particles (*e.g.* electrons, protons) behave with wave properties, exhibiting interference and diffraction. Conversely, phenomena traditionally thought of as waves (*e.g.* electromagnetic waves) behave with particle properties, carrying momentum and discrete quantities of energy. A wider discussion of this phenomenon is outwith the scope of this text; however, it can be succinctly summarised as follows:

The behaviour of any quantum mechanical particle can be fully described by a wave.

The essence of chemistry is the behaviour of electrons: how they behave and how they move from one molecule to another in a chemical reaction. To understand the electron—and hence our re-actions—we must be able to handle the wave equations.

9.1.3   The Wave Function

Since the wave function is a mathematical description of an electron (or any quantum mechanical particle) from which we can determine any observable quantity, some of these observable quantities may be negative. Some of our calculations occasionally require us to find the square of a wave function; the consequence is that, at some point, we may need to take the square root of a negative value to find the wave function. We must therefore be prepared to meet complex numbers in our exploration of electrons in chemistry. As a fortunate coincidence, it turns out that by considering all wave functions as complex numbers, it greatly simplifies the mathematics. (You will see how the simplification is done towards the end of this chapter.)

9.2 Introduction to Complex Numbers

To be able to handle complex numbers in chemistry, we will need to go over what a complex number is and how to use them in the mathematical operations with which you are already familiar. Once we have addressed the concepts we need, we can then apply them to solve the Schrödinger equation.

9.2.1   What Is a 'Complex' Number?

The term 'complex' is often laden with confusion, often taken to mean 'difficult', or 'challenging'. However, in the context of a 'complex number', the term is much more familiar to us as chemists. A 'complex' in chemistry is a metal compound consisting of multiple parts, and so it is with a complex number. Here it is a number which consists of two parts, each of which can describe different parts of a system (eqn (9.2)). In reality, the term 'compound number' is perhaps more appropriate, as it eliminates the confusion with the term 'complex'.

The general form of a complex number z is:

$$z = a + ib \qquad (9.2)$$

The symbol z is the general term to indicate a complex number; it has two components, a 'real' component a and an 'imaginary' component b. To understand the origin of each of these components, we will need to use some fundamental mathematics: namely quadratic equations. You can find details about quadratic equations and how to solve them in Toolkit A.13.

If we consider a quadratic equation, such as $y = x^2 - 6x + 11$, we find the equation cannot be factorised. We could apply the quadratic formula (eqn (9.3)):

$i = \sqrt{-1}$, see Section 9.2.2.

11 is a prime number and so its only factors are 1 are 11; these cannot be added (or subtracted) to get −6 and so the equation doesn't factorise.

> This formula will always solve a quadratic equation; if you can't easily factorise it, just use this equation to solve, giving two roots.

$$x = \frac{-b \pm \sqrt{b^2 - 4ac}}{2a} \tag{9.3}$$

$$= \frac{-(-6) \pm \sqrt{(-6)^2 - 4 \times (1 \times 11)}}{2 \times (1)}$$

$$= \frac{6 \pm \sqrt{36 - 44}}{2}$$

$$= \frac{6 \pm \sqrt{-8}}{2}$$

$$= 3 \pm \sqrt{-2} \tag{9.4}$$

However, we can immediately see the problem; we have a *square root of a negative number*. Indeed, if we sketch a graph (Figure 9.1), we see that the parabola does not cross the x-axis at all, which would suggest that there are no roots to the equation. However, in mathematics, there is the *fundamental theorem of algebra* which, when paraphrased, states that *every quadratic must have two roots*—even those which do not appear to cross the x-axis when plotted![†]

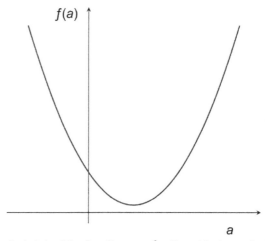

Figure 9.1   A sketch of the function $y = x^2 - 6x + 11$ shows that it does not cross the x-axis, and so does not have any 'real' roots.

9.2.2 The Square Root of a Negative Number

Overall, this presents us with a problem. For even the simplest of quadratics, $x^2 + 1 = 0$, we must accept that there are two solutions. However, to find these, we must follow the logical steps in eqn (9.5):

[†]More generally, the fundamental theorem of algebra states that any polynomial must have as many roots as its highest power; a cubic (x^3) must have three roots, a quartic (x^4) must have four, and so on.

$$x^2 + 1 = 0$$

$$x^2 = -1$$

$$x = \pm\sqrt{-1} \qquad (9.5)$$

For each root of the equation, we are led to the inexorable result that there exists a number which, when squared, is equal to -1 (eqn (9.6)). This number is a new class of numbers, which carries the label i (or, in some physics and engineering textbooks, j).

$$x^2 = -1$$

$$x = \pm i$$

$$i^2 = -1 \qquad (9.6)$$

These numbers have the spurious label of 'imaginary' numbers; however, to think of them as simply 'a figment of the imagination' is to deny their utility and versatility! This definition, however, allows us to find the square root of any negative number, an example of which is shown in eqn (9.7).

$$b^2 = -4$$

$$b = \pm\sqrt{-4}$$

$$= \pm\sqrt{4} \times \sqrt{-1}$$

$$= \pm 2 \times i \qquad (9.7)$$

$$= \pm 2i$$

> The letter "j" is used to represent the square root of -1 in a variety of programming languages, including Python, while other languages (including Matlab®) will use the letter "i".

❗ Key Concept 9.1 'Real', 'Imaginary' and 'Complex' Numbers

There are two distinct types of number; the *'real'* numbers (these are the normal 'counting' numbers with which you are already familiar) and the *'imaginary'* numbers to which you have just been introduced which, when squared, yield negative 'real' numbers. A complex number ('compound number') is a number which has both a 'real' and an 'imaginary' component:

Complex number	$\mathbb{C}$	$2 + 3i$
'Real' component	$\mathbb{R}$	2
'Imaginary' component	$\mathbb{I}$	$3i$

Important: The terms *'real'* and *'imaginary'* are nothing more than labels. Neither is any more or less 'realistic' than the other, nor is it any less valid. Some will claim that the number i is a 'pretend' number—were this to be true, it would not be as useful as it is![‡]

> The values of a and b in $z = a + bi$ are not limited to positive integer values but can take on any positive or negative value.

[‡]Historically, negative numbers were seen as 'pretend numbers', as there was no way you could have 'minus six apples'; however, they became vital when considering financial transactions for indicating debt (negative money) *versus* credit (positive money).

Example 9.1 Complex Roots for Quadratic Equations

Let's revisit a quadratic equation, $x^2 - 4x + 13 = 0$, and attempt to find the roots. We will consider this using the general quadratic formulation $ax^2 + bx + c = 0$.

- Because the value of c is a prime number (13), it has no factors other than itself and 1, and these do not add together to give $b = -4$. Therefore, we cannot factorise this. Let's instead attempt to complete the square.
- Completing the square results in the relation $(x - 2)^2 + 9 = 0$.
- This results in the relation $(x - 2) = \sqrt{-9} = \sqrt{9} \times \sqrt{-1} = \pm 3i$.
- The solution to this is therefore $(x - 2) = \pm 3i$, or $x = 2 \pm 3i$.

❓Exercise 9.1 Exercises in Computing Complex Numbers

Determine the following complex numbers in the form $a + bi$. There is no need to use decimals; expressing as $x = \sqrt{2}$ (called surds) is preferable to $x = 1.41$ .... For questions 5–8, solve the equation to find the complex number.

1. $\sqrt{-9}$
2. $\sqrt{-4}$
3. $\sqrt{-6}$
4. $\sqrt{64} - \sqrt{-36}$

5. $\left(3x^2 + 1\right) = 0$
6. $a^2 + 5 = -59$
7. $\left(b - 4\right)^2 + 26 = 1$
8. $\left(c + 2\right)^2 + 34 = -2$

9.3 Arithmetic With Complex Numbers

Arithmetic with complex numbers is straightforward; we just need to remember three key principles:

1. Add or subtract the 'real' and 'imaginary' components of the complex number separately. (We used the same idea when adding vectors in Chapter 6.)
2. When multiplying, remember that $i^2 = -1$, with the effect that multiplying two 'imaginary' numbers together gives a 'real' number.
3. When writing a fraction, avoid having a complex component as a denominator; this is carried out by using the *complex conjugate* (see Key Concept 9.2).

9.3.1 Addition and Subtraction of Complex Numbers

When *adding* or *subtracting* two complex numbers, we must add the 'real' components together to give the 'real part' of the new complex number, and then add the 'imaginary' components together to give the 'imaginary part' of the new complex number:

$$(2+3i)+(4-5i)=(2+4)+(3-5)i$$
$$=6-2i$$

> This is the same principle we followed when adding vectors; see Section 6.3.

9.3.2 Multiplication of Complex Numbers

When *multiplying* two complex numbers, we treat them the same way as when multiplying out any other algebraic example; *except* that we must remember to keep track of the i terms in accordance with the definition $i^2 = -1$:

$$(2+3i)\times(4-5i)=(2\times4)+(2\times[-5i])+(3i\times4)+(3i\times[-5i])$$
$$=(8)+(-10i)+(12i)+(-15i^2)$$
$$=(8)+(2i)+(-15\times-1)$$
$$=(8+15)+2i$$
$$=23+2i$$

❓Exercise 9.2 Handling Complex Numbers

Evaluate the following arithmetic expressions to form a new complex number in the form $a + bi$.

1. $(9 + 5i) + (8 + 5i)$
2. $(-5 + 4i) - (1 + 5i)$
3. $(-1 - 2i) \times (7 + 9i)$
4. $(-2 + 6i) + (7 + 9i)$

5. $(4 + 2i) - (2 - i)$
6. $(-5 - 5i) \times (3 - 3i)$
7. $(1 + 2i) \times (3 - 4i) + (2 - i)$
8. $(1 + i) - (2 - i) \times (1 + 3i)$

Simplify the following:

9. i^3
10. i^9
11. $3i + \sqrt{-2}$
12. $i(4 - i + 2i^2 - 3i^3)$
13. $\left(-1/2 + \sqrt{3}/2\, i\right)^3$

14. $i(3 + 2i^5)$
15. $(1-i)^2 + 3\left(2 - 1/2\, i\right)^2$
16. $(2 + 3i)(2 - 3i)$
17. $(6 + 8i)(6 - 8i)$
18. $(9 + 5i)(9 - 5i)$

> Remember that you should follow the same BIDMAS/BODMAS rules as you would otherwise, see Toolkit A.1.

9.3.3 Division of Complex Numbers

Division is slightly trickier, as we are required to eliminate the 'imaginary' term from the denominator. We do this using a device called the *complex conjugate*; this is covered in Key Concept 9.2.

> **❶ Key Concept 9.2 The complex conjugate, z^***
>
> Frequently, we wish to eliminate a complex number from one part of our equation; to do this, we use the *complex conjugate*. For any given complex number, z, the complex conjugate, z^* may be found as:
>
> $$z = a + bi$$
> $$z^* = a - bi \tag{9.8}$$
>
> When we multiply a complex number by its complex conjugate, the result we obtain is a *'real' number*:
>
> $$z \times z^* = (a + bi)(a - bi)$$
> $$= a^2 + abi + (b \times -b)(i^2) - abi$$
> $$= a^2 + abi + (-b^2)(-1) - abi \tag{9.9}$$
> $$= a^2 + b^2$$

Toolkit A.2 provides a refresher about how to correctly multiply out brackets.

Consider the fraction $f = 3/(4 + 5i)$; as with all fractions, we want to express this with only whole 'real' numbers as the denominator.

1. To eliminate the complex number from the bottom of the fraction, multiply both the top and bottom of the fraction by the *complex conjugate* (this is essentially just multiplying it by 1).

Multiplying top and bottom of a fraction by the same value is just like multiplying by 1.

If $z = 4 + 5i$, then $z^* = 4 - 5i$, so:

$$f = \frac{3}{4 + 5i} \times \frac{(4 - 5i)}{(4 - 5i)}$$

2. We then multiply out the brackets, top and bottom:

$$f = \frac{12 - 15i}{16 + (-25)(-1)} = \frac{12 - 15i}{41}$$

3. Cancel elements in the fraction if this is possible. 41 is a prime number, so cancelling here is not possible.

❓Exercise 9.3 Identifying Complex Conjugates

For each of the complex numbers, z, given, identify the complex conjugate and the value of $(z \times z^*)$.

1. $z = 4 + 2i$ 3. $z = -5 - 5i$

2. $z = 7 + 2i$ 4. $z = 3 - 3i$

5. $z = 2 - 4i$ 7. $z = 5i$

6. $z = -3 + 2i$ 8. $z = 2 - 3i$

? Exercise 9.4 Dividing Complex Numbers

For each of the division expressions, f, given, identify the appropriate complex conjugate and use it to eliminate the complex number in the denominator.

1. $f = \dfrac{-1}{(-5 + 2i)}$ 5. $f = \dfrac{(3 + 7i)}{(-1 - i)}$

2. $f = \dfrac{9}{(7 - 2i)}$ 6. $f = \dfrac{(1 + 6i)}{(-2 + 4i)}$

3. $f = \dfrac{6i}{(2 - 5i)}$ 7. $f = \dfrac{(2 + 5i)}{(-2 - 3i)}$

4. $f = \dfrac{-5i}{(2 - i)}$ 8. $f = \dfrac{(7 + 9i)}{(-2 - 5i)}$

9.4   Alternative Representations of Complex Numbers

9.4.1   Graphical Representations

Since a complex number consists of *two independent—but inextricably linked—components*, we have another way to describe these numbers. Complex numbers may be plotted on a graph, where the 'real' component is plotted on one axis (the 'x'-axis) and the 'imaginary' component is plotted on the other axis (the 'y'-axis). This is called an *Argand diagram*, and an example is shown in Figure 9.2.

This allows us to define a complex number in terms of a *modulus* (radial distance from the origin) and an *argument* (angle from the 'real' axis); you will recognise this representation from polar coordinates (Chapter 5), and can see that the complex conjugate, z^*, of a complex number, z, is just a 'reflection' over the 'real axis' of z's position on the Argand diagram. Key Concept 9.3 shows how to convert between the different notations.

This shows how a complex number can be used to represent two quantities which are orthogonal to each other (a 'real' quantity and an 'imaginary' quantity). This means that complex numbers are used extensively in electromagnetism and optics (where the electric and magnetic fields of light are orthogonal), in electronics (particularly with alternating (AC) current) and in fluid dynamics (really useful if looking at flows through junctions!). However, to chemists, the most important use is principally in quantum mechanics and wave notation for either a wave–particle or wave functions.

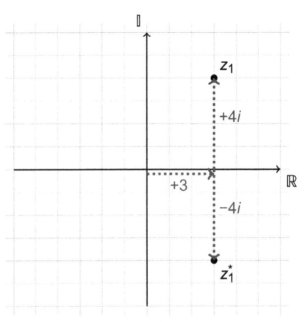

(a) $z = a + ib$, $z^* = a - ib$.

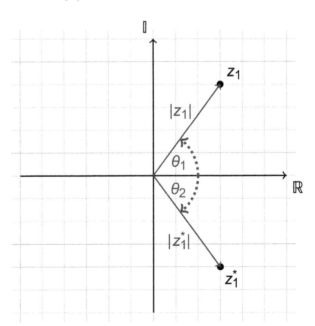

(b) $z = |z|, \theta$, $z^* = |z|, -\theta$.

Figure 9.2 Visual representations of the complex number $z_1 = 3 + 4i$, and its complex conjugate $z_1^* = 3 - 4i$ in $a + ib$ form and as a modulus and argument, $|z|, \theta$. Representing complex numbers in a geometric manner such as this can unlock fresh understanding about the nature of complex numbers and how they interact with each other. Note the axis labels for the 'real' and 'imaginary' directions.

> ### ❶ Key Concept 9.3 Converting from *a + bi* to |z|, θ
>
> As we saw in Chapter 5, we can convert readily between coordinate systems. For complex numbers, the principle is much the same.
>
> For the complex number $z = a + bi$:
>
> $$|z|^2 = a^2 + b^2$$
> $$|z| = \sqrt{a^2 + b^2}$$
>
> You will recognise this from Pythagoras' theorem. If we are considering triangular geometry, there is also a relationship with the angle θ:
>
> $$\cos\theta = \frac{a}{|z|}$$
> $$\sin\theta = \frac{b}{|z|}$$
> $$\theta = \cos^{-1}\left(\frac{a}{|z|}\right) = \sin^{-1}\left(\frac{b}{|z|}\right)$$
>
> This is just applying the trigonometric functions from Chapter 5 to Figure 9.2.
>
> For the complex number $z = |z|$, θ:
>
> $$a = |z|\cos\theta$$
> $$bi = i|z|\sin\theta$$
>
> Once again, this is applying trigonometric rules. The whole complex number can then be represented as:
>
> $$z = a + bi = |z|(\cos\theta + i\sin\theta)$$
>
> This will be an important result which we will further develop later.

❓ Exercise 9.5 Converting Between Complex Identities

For each of the following complex numbers, written in the form $z = a + bi$, calculate the modulus of the complex number, $|z|$ and its principal argument, θ, in radians. Remember that the argument should lie between $-\pi$ and π.

> When you calculate the angle, it is worth sketching a diagram to ensure you are getting the angle you expect; there are multiple possible angles.

1. $3 + i$
2. $-6 - 6i$
3. $-3 - 9i$
4. $-9 + 0i$
5. $4 + 3i$
6. $0 + 6i$
7. $-9 - 6i$
8. $-4 + 6i$

9.4.2 Exponential Representations

A final way to represent complex numbers is as an exponential. This representation shows a complex number in the form $z = Ae^{i\theta}$. Whilst eqn (9.10) seems an unusual way to represent the complex number, it can be shown that these two statements are equivalent using the Maclaurin expansion for $e^{i\theta}$ (see Toolkit A.12 for the expansions of e^{Ax}, $\sin(Bx)$ and $\cos(Cx)$):

$$z_1 = Ae^{i\theta}$$
$$= A\left(\cos\theta + i\sin\theta\right) \tag{9.10}$$

▪■
$e^{i\theta} = \cos\theta + i\sin\theta$

This geometric connection also offers a means to simplify any instances of raising a complex number to a power:

$$z_2 = A(e^{i\theta})^k = A(e^{ik\theta})$$
$$= A(\cos\theta + i\sin\theta)^k = A(\cos(k\theta) + i\sin(k\theta)) \tag{9.11}$$

where k is simply a constant.

Eqn (9.11) is *De Moivre's theorem* and by using this we can determine the result of raising a complex number to any power. While we are accustomed to any relation such as e^{kx} as a runaway exponential describing exponential growth, by including a complex number, we have created a pair of interlinked cyclic oscillating functions (the sine and cosine functions), which describe a wave.

In Key Concept 9.3, we showed that any complex number can be converted to a sine or cosine representation, as:

$$z = a + ib$$
$$= |z|\left(\cos\theta + i\sin\theta\right) \tag{9.12}$$

The application of De Moivre's theorem to manage the indices of complex numbers is shown in Example 9.2.

Example 9.2 Simplifying Exponents of Complex Numbers

Consider the complex number

$$z^{16} = (1 - i)^{16} \tag{9.13}$$

We could look to solve it by multiplying out the brackets:

$$(1 - i) \times (1 - i) \times (1 - i) \times (1 - i) \times (1 - i) \times (1 - i) \times (1 - i) \times \cdots$$

which is tedious to say the least; however, lets use De Moivre's theorem to simplify this.

• Firstly, determine the cosine or sine equivalent complex number by evaluating $|z|$ and θ:

$$|z| = \sqrt{a^2 + b^2} \qquad \theta = \sin^{-1}\left(\frac{b}{|z|}\right)$$

$$= \sqrt{1^2 + (-1)^2} \qquad = \sin^{-1}\left(\frac{-1}{\sqrt{2}}\right)$$

$$= \sqrt{2} \qquad = -0.785\,\text{rad} \equiv -\frac{\pi}{4}$$

- Now write the complex number in terms of the modulus and its argument:

$$z = |z|(\cos\theta + i\sin\theta)$$
$$= \sqrt{2}\left(\cos\left[-\frac{\pi}{4}\right] + i\sin\left[-\frac{\pi}{4}\right]\right)$$

- We now use De Moivre's theorem to express this as an exponent:

$$z = |z|e^{i\theta}$$
$$= \sqrt{2}\,e^{\left[-i\frac{\pi}{4}\right]}$$

- Finally, we can now look at the exponent, z^{16}, and use the same rules of indices as described in Chapter 1:

$$z^{16} = \left(\sqrt{2}\,e^{-\frac{i\pi}{4}}\right)^{16}$$
$$= \left(\sqrt{2}\right)^{16} \times e^{\left[16 \times -\frac{i\pi}{4}\right]} \qquad (9.14)$$
$$= 2^8 \times e^{-4i\pi}$$

- Should we desire, we can then translate this back to the $z = a + ib$ format, using the principles shown in Key Concept 9.3:

$$(1-i)^{16} = 2^8 \times e^{-4i\pi}$$
$$= 2^8 \times \left[\cos(-4\pi) + i\,\sin(-4\pi)\right]$$
$$= 256 \times 1 + 256 \times 0i \qquad (9.15)$$
$$= 256$$

Given De Moivre's theorem and some simple rules of indices, raising a complex number to an index becomes a straightforward way of dealing with powers of complex numbers.

❓Exercise 9.6 Exponential Notation of Complex Numbers

Sketch the following complex numbers on an Argand diagram and express using exponential notation:

1. $z = 1 + 2i$
2. $z = 1 - i$
3. $z = 3 + 4i$
4. $z = 2 - 0.5i$
5. $z = 3 - i$
6. $z = 2 + 2i$
7. $z = -8 - 5i$
8. $z = -1 - 9i$
9. $z = -3 - 4i$
10. $z = 9 + 10i$

? Exercise 9.7 Dealing with Indices in Complex Numbers

Remember that the argument (angle) must be determined in radians.

For each of the following, determine:

1. $z = (1 + 2i)^9$

2. $z = (1 - i)^{23}$

3. $z = (3 + 4i)^{13}$

4. $z = (2 - 0.5i)^5$

5. $z = (3 - i)^{-3}$

6. $z = (2 + 2i)^{-7}$

7. $z = (-8 - 5i)^{-2}$

8. $z = (-1 - 9i)^{-7}$

9. $z = (-3 - 4i)^0$

10. $z = (9 + 10i)^{-5}$

9.5 Complex Numbers and Wave Functions

All of the mathematical development shown so far leads us to the conclusion that there is an intimate connection between complex numbers and trigonometric functions. You will recall from Chapter 5 that a graph of either $\cos(x)$ or $\sin(x)$ shows a wave, and it is this which makes complex numbers so useful for describing a wave function; the two interlinked components of e^{ikx} are perfect for describing an electromagnetic wave with its interlinked electric and magnetic fields.

9.5.1 Models in Quantum Mechanics

A fundamental model in quantum mechanics is the 'particle in a box'. An electron (or other quantum mechanical particle) is restricted to a defined space of zero potential energy. It can freely move within this space, but cannot leave the space and its only energy is kinetic energy. The simplest version of this problem is the one-dimensional box; *i.e.* the electron can only move in a single dimension, usually defined as the x-axis.

The general form of the wave function for this 'bounded particle' $\Psi_{\text{B.P.}}$ is shown in eqn (9.16) and is obtained by combining the 'free particle' wave function in the $+x$ direction with that of the wave function in the $-x$ direction (the particle travelling in the $+x$ direction, reflecting from the barrier and coming back in the $-x$ direction, reflecting again and going back in the $+x$ direction and so on)

$$\Psi_{\text{B.P.}} = \Psi_{+x} \pm \Psi_{-x}$$
$$= Ae^{ikx} \pm Ae^{-ikx} \tag{9.16}$$

We now know that we can use De Moivre's theorem to identify viable wave functions: either:

$$\Psi_{\text{B.P.}} = Ae^{ikx} + Ae^{-ikx}$$
$$= A\left(\left[\cos(kx) + i\sin(kx)\right] + \left[\cos(kx) - i\sin(kx)\right]\right)$$
$$= A(2\cos(kx))$$

or

$$\Psi_{\text{B.P.}} = Ae^{ikx} - Ae^{-ikx}$$
$$= A\left(\left[\cos(kx) + \text{i}\sin(kx)\right] - \left[\cos(kx) - \text{i}\sin(kx)\right]\right)$$
$$= A\left(2\text{i}\sin(kx)\right)$$

For the particle in a box, the wave function must be zero at the edges of the box, which means that the 'real' resultant wave function $(2A\cos(kx))$ cannot fit as $\cos(0) = 1$. This means that the wave function we are interested in is actually the 'imaginary' component, $(2A\text{i}\sin(kx))$. This highlights the trouble with the word 'imaginary' in the context of complex numbers—the wave function is absolutely valid; it is simply that we are interested in the part of the wave function which is labelled with i.

> This is a *boundary condition*; we listed these in Section 8.6.

Now that we understand how complex numbers work, we can apply them to solving the Schrödinger equation.

9.5.2 Solving the Schrödinger Equation

We have identified the form of a wave function as a complex number, and the simplest form to use for finding solutions to the Schrödinger equation is its exponential form, $\Psi_x = Ae^{ikx}$. Since the wave function is a mathematical description of all aspects of the behaviour of a quantum particle, to get useful information from the wave function we use an appropriate operator. A list of some operators is shown in Table 9.1.

> We discussed the values of A and k for the particle in a one-dimensional box in Section 8.6; the normalisation and determination of these values will only make sense in the context of the selection rules and using De Moivre's theorem to express the wave function as a sum of sine or cosine components.

Table 9.1 A list of quantum mechanical operators. Note the (x) showing operators in a single dimension, while others are over all space (x, y, z).

Operator	Symbol	Operation
Momentum (x)	$\hat{\mathbf{p}}$	$-\text{i}\hbar\dfrac{\partial}{\partial x}$
Kinetic energy (x)	$\hat{\mathbf{T}}_x$	$-\dfrac{\hbar^2}{2m}\dfrac{\partial^2}{\partial x^2}$
Position (between $x = a$ and $x = b$)	$\hat{\mathbf{x}}$	$\displaystyle\int_a^b \Psi\Psi^*\partial x$
Potential energy	$\hat{\mathbf{V}}$	$V(r)$
Kinetic energy (x, y, z)	$\hat{\mathbf{T}}$	$-\dfrac{\hbar^2}{2m}\left[\dfrac{\partial^2}{\partial x^2} + \dfrac{\partial^2}{\partial y^2} + \dfrac{\partial^2}{\partial z^2}\right]$
Total energy (x, y, z)	$\hat{\mathbf{E}}$	$-\dfrac{\hbar^2}{2m}\left[\dfrac{\partial^2}{\partial x^2} + \dfrac{\partial^2}{\partial y^2} + \dfrac{\partial^2}{\partial z^2}\right] + V(r)$
		$-\dfrac{\hbar^2}{2m}\nabla^2 + V(r)$

> The operators in this table are for the time-independent Schrödinger equation, allowing determination of properties as they vary with position rather than time. A separate set of operators exist for the time-dependent Schrödinger equations.

You will notice that all operators rely on calculus, considering either a rate of change of the wave function or an area enclosed by the square of the wave function (formally $\Psi \times \Psi^*$ to give a 'real' result). The application of these operators in one dimension is trivial, however it is essential in computational chemistry to calculate the observable property (*i.e.* the property which we would measure in an experiment) to compare measurements with predictions. The application of this to a particle in one dimension is shown in Example 9.3.

Example 9.3 Applying Operators to a Complex Wave Function

Q: The behaviour of a free electron moving in one dimension may be described by the wave function

$$\Psi_x = Be^{(ikx)}$$

Determine an expression to quantify the momentum of a free electron in terms of the constants B and k.

A: First of all, we need to identify the appropriate operator from Table 9.1; we see that the operator we need is $\hat{p}x = -i\hbar \dfrac{\partial}{\partial x}$.

• Writing out the Schrödinger equation in full, we obtain the form:

$$\hat{H}\Psi = E\Psi$$

$$-i\hbar \frac{\partial \Psi}{\partial x} = -i\hbar \frac{\partial \left(Be^{(ikx)}\right)}{\partial x} = E\Psi = E \times \left(Be^{(ikx)}\right)$$

• We need to differentiate our expression (Chapter 7). Remember, i is a constant and we treat it as any other:

$$-i\hbar \frac{\partial \left(Be^{(ikx)}\right)}{\partial x} = -i\hbar \left[ik \times Be^{ikx}\right]$$

$$= -i^2 \hbar k Be^{(ikx)}$$

$$= -1 \times -1 \times \hbar k Be^{(ikx)}$$

$$= \hbar k Be^{(ikx)}$$

• Now that we have our first derivative, we need to see if we have returned the original wave function; we can see that this has been returned (the benefit of using the exponent notation of our wave function!):

$$-i\hbar\frac{\partial\left(Be^{(ikx)}\right)}{\partial x} = \hbar k Be^{(ikx)}$$

$$= \hbar k \times \Psi_x$$

- This means that our eigenvalue, E, from the Schrödinger equation is given by the expression:

$$E = \hbar k$$

and this quantifies the momentum of a free electron. Note that the final result is independent of position; it is a constant value, showing that a free electron carries constant momentum.

9.6 Summary

In this chapter, we have introduced the complex number; a type of number which arises from the need to have fully reversible mathematics (*i.e.* being able to perform every type of operation on every number). In this case, we needed to be able to apply a square root to a negative number. We have also used the discussion of complex numbers to tie together a number of concepts from across the book; namely calculus, coordinate geometry and trigonometry, as well as applying our knowledge from vectors on how to combine orthogonal components.

In chemistry, complex numbers are encountered in quantum mechanics when describing two orthogonal but linked properties of a quantum wave–particle. They are not widely used elsewhere in chemistry; however, they are a crucial part of computational chemistry, an ever-growing field of research and application.

Summary: Chapter 9

- Complex numbers arise from the solution of the identity $i = \sqrt{-1}$. The definition of i is given as:

$$i^2 = -1$$

- A complex number is a number which has a 'real' component, a, and an 'imaginary' component, bi, and has the general form:

$$z = a + bi$$

- This can be plotted geometrically on an Argand diagram with a modulus $|z| = \sqrt{a^2 + b^2}$ and an argument $\theta = \sin^{-1}\left(\frac{b}{|z|}\right)$. This allows a complex number to be written in trigonometric form:

$$z = |z|(\cos\theta + i\sin\theta)$$

- This, in turn, may be written in an exponential form, *via* De Moivre's theorem:

$$z = |z|e^{i\theta}$$

- Complex numbers can be added and subtracted by considering the real and imaginary components separately; they can be multiplied as with any other set of brackets, remembering the identity $i^2 = -1$.

- Division of complex numbers requires the complex conjugate, z^*. The complex conjugate is a number which when multiplied by its originating complex number, returns a fully real result:

$$z = a \pm bi$$
$$z^* = a \mp bi$$
$$zz^* = a^2 + b^2$$

- Complex numbers can then be divided by multiplying both top and bottom of the fraction by the complex conjugate of the denominator; this eliminates the complex component from the bottom of the fraction.

- Expressing a complex number as an exponential can greatly simplify the process of calculus and other mathematical operations.

?Exercise 9.8 Complex Numbers in a Chemical Context

In this exercise, you will use operators listed in Table 9.1 on the complex wave functions given. You will also need to use techniques covered in other chapters (particularly Chapters 7 and 8).

1. Determine the time-independent momentum and kinetic energy of a wave–particle whose wave function is given by $\Psi = Ae^{2\pi ix/\lambda}\,e^{-2\pi i\nu t}$, where λ and ν correspond to the wavelength and frequency of the particle wave, respectively.
2. Determine the momentum and kinetic energy of a wave–particle whose wave function is given by $\Psi_r = Be^{i\pi\nu r}$, where ν corresponds to the frequency of the particle wave.
3. Determine the complex conjugate of the wave function $\Psi_r = Be^{i\pi\nu r}$ and evaluate the expression $\Psi_r\Psi_r^*$.

Solutions: Exercise

Solutions: Exercise 9.1

1. $\pm 3i$

5. $x = \pm \dfrac{i}{\sqrt{3}}$ or $\pm \dfrac{\sqrt{3}}{3}i$

2. $\pm 2i$

6. $a = \pm 8i$

3. $\pm\sqrt{6}i$

7. $b = 4 \pm 5i$

4. $\pm 8 \mp 6i$

8. $c = -2 \pm 6i$

Solutions: Exercise 9.2

1. $17 + 10i$

10. i

2. $-6 - i$

11. $\left(3 + \sqrt{2}\right)i$

3. $11 - 23i$

12. $-2 + 2i$

4. $5 + 15i$

13. 1

5. $2 + 3i$

14. $-2 + 3i$

6. -30

15. $\dfrac{45}{4} - 8i$

7. $13 + i$

16. 13

8. $-4 - 4i$

17. 100

9. $-i$

18. 106

Solutions: Exercise 9.3

1. $z^* = 4 - 2i,\ zz^* = 20$

5. $z^* = 2 + 4i,\ zz^* = 20$

2. $z^* = 7 - 2i,\ zz^* = 53$

6. $z^* = -3 - 2i,\ zz^* = 13$

3. $z^* = -5 + 5i,\ zz^* = 50$

7. $z^* = -5i,\ zz^* = 25$

4. $z^* = 3 + 3i,\ zz^* = 18$

8. $z^* = 2 + 3i,\ zz^* = 13$

Solutions: Exercise 9.4

1. $\dfrac{z_1}{z_2} = \dfrac{5}{29} + \dfrac{2}{29}i;\ z_2^* = -5 - 2i$

5. $\dfrac{z_1}{z_2} = -5 - 2i;\ z_2^* = -1 + i$

2. $\dfrac{z_1}{z_2} = \dfrac{63}{53} + \dfrac{18}{53}i;\ z_2^* = 7 + 2i$

6. $\dfrac{z_1}{z_2} = \dfrac{11}{10} - \dfrac{4}{5}i;\ z_2^* = -2 - 4i$

3. $\dfrac{z_1}{z_2} = -\dfrac{30}{29} + \dfrac{12}{29}i;\ z_2^* = 2 + 5i$

7. $\dfrac{z_1}{z_2} = -\dfrac{19}{13} - \dfrac{4}{13}i;\ z_2^* = -2 + 3i$

4. $\dfrac{z_1}{z_2} = 1 - 2i;\ z_2^* = 2 + i$

8. $\dfrac{z_1}{z_2} = -\dfrac{59}{29} + \dfrac{17}{29}i;\ z_2^* = -2 + 5i$

Solutions: Exercise 9.5

1. $|z| = \sqrt{10}; \theta = \arctan\left(\dfrac{1}{3}\right) \equiv 0.32^c$

2. $|z| = 3\sqrt{8}; \theta = \arctan\left(\dfrac{-6}{-6}\right) \equiv -2.36^c$

3. $|z| = 3\sqrt{10}; \theta = \arctan\left(\dfrac{-9}{-3}\right) \equiv -1.89^c$

4. $|z| = 9; \theta = \arctan\left(\dfrac{0}{-9}\right) \equiv 0^c$

5. $|z| = 5; \theta = \arctan\left(\dfrac{3}{4}\right) \equiv 0.64^c$

6. $|z| = 6; \theta = \arctan\left(\dfrac{6}{0}\right) \equiv 1.57^{c*}$

7. $|z| = 3\sqrt{13}; \theta = \arctan\left(\dfrac{-6}{-9}\right) \equiv -2.55^c$

8. $|z| = 2\sqrt{13}; \theta = \arctan\left(\dfrac{6}{-4}\right) \equiv 2.16^c$

*The arctan for Question 6 is an impossible result; you will need a diagram to find the angle, $\pi/2$

Solutions: Exercise 9.6

Remember sketches only need to show the key aspects of a graph, including shape and key characteristics; they are there to help with visualisation.

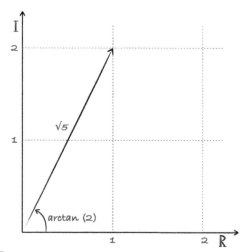

1. $z = \sqrt{5}\,e^{1.11i}$

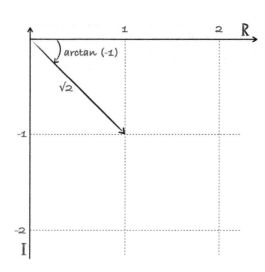

2. $z = \sqrt{2}\, e^{-0.79i}$

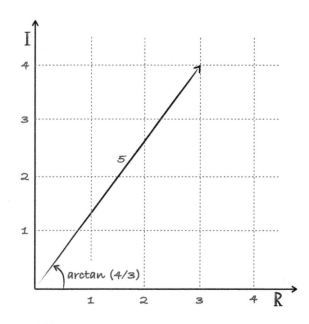

3. $z = 5\, e^{0.93i}$

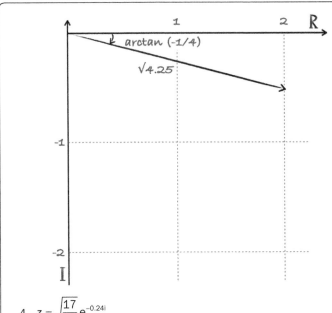

4. $z = \sqrt{\dfrac{17}{4}}\, e^{-0.24i}$

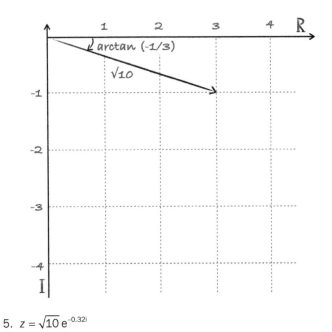

5. $z = \sqrt{10}\, e^{-0.32i}$

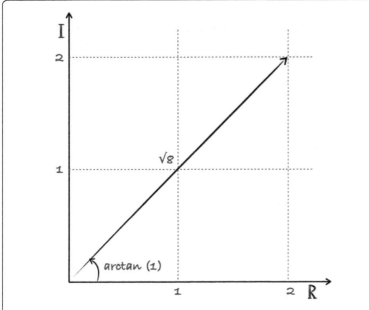

6. $z = \sqrt{8}\,e^{0.79i}$

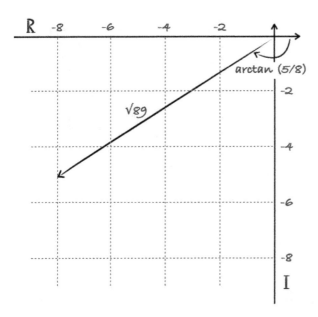

7. $z = \sqrt{89}\,e^{-2.58i}$

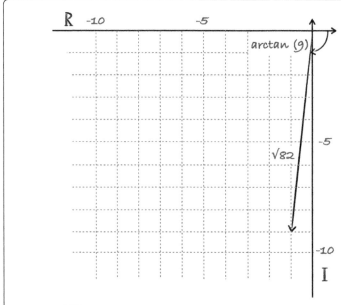

8. $z = \sqrt{82}\, e^{-1.68i}$

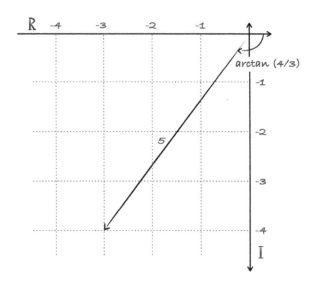

9. $z = 5 e^{-2.21i}$

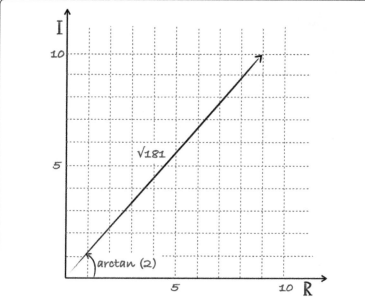

10. $z = \sqrt{181}\, e^{0.84i}$

Solutions: Exercise 9.7

1. $z = \left(\sqrt{5}\right)^9 \times e^{9 \times 1.11i}$
   $\equiv \left(1.40 \times 10^3\right) e^{9.99i}$
   $\equiv -\left(1.20 \times 10^3\right)$
   $\quad -\left(7.18 \times 10^2\right)i$

4. $z = \left(\sqrt{\dfrac{17}{4}}\right)^5 \times e^{5 \times -0.24i\,i}$
   $\equiv \left(3.72 \times 10^1\right) e^{-1.2i}$
   $\equiv \left(1.26 \times 10^1\right)$
   $\quad -\left(3.50 \times 10^1\right)i$

2. $z = \left(\sqrt{2}\right)^{23} \times e^{23 \times -0.79i}$
   $\equiv \left(2.90 \times 10^3\right) e^{-18.17i}$
   $\equiv -\left(2.05 \times 10^3\right)$
   $\quad +\left(2.05 \times 10^3\right)i$

5. $z = \left(\sqrt{10}\right)^{-3} \times e^{-3 \times -0.32i}$
   $\equiv \left(3.16 \times 10^{-2}\right) e^{0.96i}$
   $\equiv \left(1.80 \times 10^{-2}\right)$
   $\quad +\left(2.60 \times 10^{-2}\right)i$

3. $z = (5)^{13} \times e^{13 \times 0.93i}$
   $\equiv \left(1.22 \times 10^9\right) e^{12.09i}$
   $\equiv \left(1.06 \times 10^9\right)$
   $\quad -\left(5.98 \times 10^8\right)i$

6. $z = \left(\sqrt{8}\right)^{-7} \times e^{-7 \times 0.79i}$
   $\equiv \left(6.91 \times 10^{-4}\right) e^{-5.53i}$
   $\equiv \left(4.88 \times 10^{-4}\right)$
   $\quad +\left(4.88 \times 10^{-4}\right)i$

7. $z = \left(\sqrt{89}\right)^{-2} \times e^{-2 \times -2.58i}$

   $\equiv \left(1.12 \times 10^{-2}\right) e^{5.16i}$

   $\equiv \left(4.92 \times 10^{-3}\right)$

   $\quad - \left(1.01 \times 10^{-2}\right) i$

8. $z = \left(\sqrt{82}\right)^{-7} \times e^{-7 \times -1.68i}$

   $\equiv \left(2.00 \times 10^{-7}\right) e^{11.76i}$

   $\equiv \left(1.40 \times 10^{-7}\right)$

   $\quad - \left(1.43 \times 10^{-7}\right) i$

9. $z = \left(5\right)^{0} \times e^{0 \times -2.21i_i}$

   $\equiv 1 e^{0i} \equiv 1$

10. $z = \left(\sqrt{181}\right)^{-5} \times e^{-5 \times 0.84i_i}$

    $\equiv \left(2.27 \times 10^{-6}\right) e^{-4.2i}$

    $\equiv -\left(1.13 \times 10^{-6}\right)$

    $\quad + \left(1.97 \times 10^{-6}\right) i$

Solutions: Exercise 9.8

1. Momentum:

$$\hat{\mathbf{p}} \times \psi = -i\hbar \left(\frac{\partial \psi}{\partial x}\right)_t = p_x \psi$$

$$= -i\hbar \cdot \frac{2\pi\, i}{\lambda} \cdot A e^{\frac{2\pi\, ix}{\lambda}} e^{-2\pi i \nu t}$$

$$= -\frac{2\pi\, i\hbar}{\lambda} \cdot \psi$$

$$p_x = -\frac{2\pi\, i^2 \hbar}{I} \equiv \frac{2\pi\, \hbar}{\lambda} = \frac{h}{\lambda}$$

Kinetic energy:

$$\hat{\mathbf{T}}_x \psi = -\frac{\hbar^2}{2m} \left(\frac{\partial^2 \psi}{\partial x^2}\right)_t = T_x \psi$$

$$= -\frac{\hbar^2}{2m} \cdot \frac{4\pi^2 i^2}{\lambda^2} \cdot A e^{\frac{2\pi\, ix}{\lambda}} e^{-2\pi i \nu t}$$

$$= -\frac{4\pi^2 i^2 \hbar^2}{2m\lambda^2} \cdot \psi$$

$$T_x = -\frac{4\pi\, i^2 \hbar^2}{2m\lambda^2} \equiv \frac{4\pi\, \hbar^2}{2m\lambda^2} \equiv \frac{h^2}{2m\lambda}$$

2. Momentum:

$$p_x = \hbar \pi \nu \equiv \frac{h\nu}{2}$$

Kinetic energy:

$$T_x = -\frac{\hbar^2}{2m} \cdot -\pi^2 \nu^2 \equiv \frac{h^2\nu^2}{8m}$$

3. Complex conjugate,

$$\psi_r^* = Be^{-i\pi\nu r^i}$$

$$\psi_r \psi_r^* = Be^{i\pi\nu r} \times Be^{-i\pi\nu r}$$
$$= B^2 \times e^{(i\pi\nu r - i\pi\nu r)}$$
$$= B^2 e^0 = B^2$$

Appendix

The Mathematical Toolkit

Throughout this text, we have drawn on mathematical knowledge which is outwith the scope of the narrative of this text; however, having it to hand as a reference can give additional depth of understanding. There are also a number of occasions when 'standard techniques' are used (including orders of operations, bracket expansion and so on), and we feel it is worth including these so that we all agree a standard set of mathematical rules.

Inclusion of these would be incongruous in the main body of the text, so we have instead included these as 'toolkit' items; maths which is useful in the context of this text, but does not belong in the main body of the work.

As with the key concepts, these are intended to be a 'dip-in' resource, written as quick reference tools. They are included in no particular order. We hope that they offer a useful addition to the content of the book.

Toolkit A.1 BIDMAS (or BODMAS, or PEMDAS)

This is also known as the 'order of operations' for prioritising mathematical operations in arithmetic calculations.

Consider a popular 'social media' problem:

$$9 + 4 - 5 + 6 + 2 - 8 \div 4 - 6 \times 3 + 11 \times 8 + 12 = ? \qquad \text{(A.1)}$$

Try to evaluate the expression *without using a calculator*; it is often useful to do this as an individual within a group and agree on an outcome. Depending on the group, you will come out with answers which will vary; one answer often produced is +4, while another answer is +96.

The BIDMAS framework (or BODMAS, BEDMAS or PEMDAS, depending on locale) gives a structure for correctly evaluating such problems.

B rackets	B rackets	P arentheses
O rder	I ndices	E xponents
D ivision	D ivision	M ultiplication
M ultiplication	M ultiplication	D ivision
A ddition	A ddition	A ddition
S ubtraction	S ubtraction	S ubtraction

This gives an instruction set for evaluating complex expressions:

1. Evaluate all bracketed expressions (expressions in parentheses).
2. Evaluate all expressions with a higher order (index, or exponential), e.g. 2^4.
3. Carry out all multiplications and divisions.[†]
4. Carry out all addition and subtractions.

There are numerous claimed counterexamples to the order of operations above; it is important to remember that division and multiplication are mutually inverse functions and as such have the same priority, as are addition and subtraction. Any division can be rewritten as a multiplication, and any subtraction can be rewritten as an addition:

$$2 \div 4 \times 5 \equiv 2 \times \frac{1}{4} \times 5 = 2.5$$

$$12 - 4 + 5 \equiv 12 + (-4) + 5 = 13$$

This eliminates any ambiguity concerning the order of mutually inverse operations and allows the correct calculation of the arithmetic.

Following this guidance, you should be able to identify the correct value for eqn (A.1). Of course, when writing such problems, it is advisable to use brackets for clarity; indeed most formulae in chemistry include brackets to eliminate such ambiguity and it is worth preserving these brackets in your arithmetic calculations.

Toolkit A.2 Expanding Brackets

To expand brackets, you should use the following logic:

$$f(x) = (ax + b)(cx + d)$$
$$= acx^2 + adx + bcx + bd$$

This arises from the following procedure:

$$(ax + b)(cx + d)$$

Operation 1 :	$ax \times cx$	=	acx^2
Operation 2 :	$b \times d$	=	bd
Operation 3 :	$b \times cx$	=	bcx
Operation 4 :	$ax \times d$	=	adx
Total sum :	$acx^2 + adx + bcx + bd$		

Remember to pay attention to the signs if any of a, b, c or d are negative:

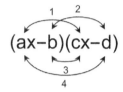

$$(ax-b)(cx-d)$$

Operation 1 :	$ax \times cx$	=	acx^2
Operation 2 :	$-b \times -d$	=	$+bd$
Operation 3 :	$-b \times cx$	=	$-bcx$
Operation 4 :	$ax \times -d$	=	$-adx$
Total sum :	$acx^2 - adx - bcx + bd$		

Where there are three brackets to multiply together (e.g. $(ax + b)(cx + d)$ $(gx + h)$), simply multiply the first two brackets together, and then multiply each term stepwise again with the third bracket.

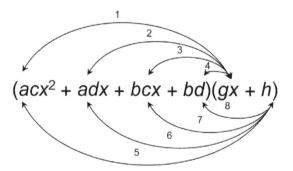

$$(acx^2 + adx + bcx + bd)(gx + h)$$

This process is extended easily for as many brackets as needed; however, it is important to keep track of your signs.

Toolkit A.3 Combining Equations

Often, we must combine a number of equations in order to eliminate a particular component. We do this simply by adding (or subtracting) one equation to (or from) another in its entirety. We see this in two main areas in chemistry:

1. *Simultaneous equations:* Consider the following pair of equations:

$$4x - y = 16$$
$$x + y = 4$$

   To eliminate the y term, we *add* the two equations together:

$$(4x - y) + (x + y) = (16 + 4)$$

   This allows us to determine that $5x = 20$, therefore $x = 4$. This can then be substituted into either equation to identify that $y = 0$. On occasions, we may need to multiply an equation by a constant to eliminate one of the variables.

2. *Redox equilibria:* In chemistry we add equations together to eliminate the electron term, allowing us to express an overall equilibrium for the redox process. Consider the following pair of reduction half-equations:

$$Na^+(aq) + e^- \rightleftharpoons Na(s)$$

$$\tfrac{1}{2}Cl_2(g) + e^- \rightleftharpoons Cl^-(aq)$$

   To eliminate the electron term, we *subtract* one equation from the other, and rearrange (as we do for mathematical equations):

$$\left(\tfrac{1}{2}Cl_2(g) + e^-\right) - \left(Na^+(aq) + e^-\right) \rightleftharpoons Cl^-(aq) - Na(s)$$

$$\tfrac{1}{2}Cl_2(g) + Na(s) \rightleftharpoons Cl^-(aq) + Na^+(aq)$$

Toolkit A.4 Powers and Exponents: Types of Index

It is important to recognise the difference between a power and an exponent as they are similar in appearance and often confused.

They have the general form:

$$b^c$$

where b is the *base* and c is the *index*.

In a *power* relationship, the *base* is the variable and is raised to a constant index. Examples of powers include:

$$y = x^2$$

$$\text{Rate} = k[N_2O_5]^2$$

$$\text{Volume} = \frac{4}{3}\pi r^3$$

A root is the *inverse* of a power relationship and can be represented by a fractional index:

$$\sqrt{y} = y^{1/2} = x$$

$$\sqrt{\frac{\text{rate}}{k}} = \left(\frac{\text{rate}}{k}\right)^{1/2} = [N_2O_5]$$

$$\sqrt[3]{\frac{3}{4\pi} \times \text{volume}} = \left(\frac{3}{4\pi} \times \text{volume}\right)^{1/3} = r$$

In an *exponential* relationship, a constant base is raised to a *variable* index. Examples of exponentials include:

$$y = 2^x$$

$$A = Ae^{-E_a/RT}$$

$$[H^+] = 4.5 \times 10^{-3} \text{ mol dm}^{-3}$$

Toolkit A.5 Combining Indices

Fundamentally, an index represents repeated multiplication of a number, in the form:

$$2^4 = 2 \times 2 \times 2 \times 2$$

With a general expression for a power being:

$$a^n = \underbrace{a \times a \times \ldots \times a}_{n}$$

When indices are combined, it can be useful to think in these terms:

$$3^2 \times 3^4 = \underbrace{3 \times 3}_{3^2} \times \underbrace{3 \times 3 \times 3 \times 3}_{3^4}$$
$$= 3^6$$
$$= 3^{2+4}$$

A similar argument may be proposed to demonstrate the division property of index combination.

This only applies when the numbers have a common base; $3^2 \times 5^8$ cannot be simplified further, as the bases are not the same.

In general, the following principle applies:

$$m^a \times m^b = m^{a+b}$$

$$\frac{p^a}{p^b} = p^{a-b}$$

Toolkit A.6 Converting Logarithms

A logarithm of one base may be converted to another base using the following steps.

1. If we want to convert $\ln(A)$ to a log base of 10, let's first say that the value is equal to x:

$$\ln(A) = \log_e(A) = x$$

2. This allows us to express A in terms of x:

$$e^x = A$$

3. We now apply our new log base of 10 to both sides and then take the x outside the logarithm:

$$\log_{10}(e^x) \equiv x\,\log_{10}(e) = \log_{10}(A)$$

4. We already have a definition of $x = \ln(A)$ from step 1, so we apply this:

$$\ln(A) \times \log_{10}(e) = \log_{10}(A)$$

5. We can now rearrange this to obtain the conversion we desire:

$$\ln(A) = \frac{\log_{10}(A)}{\log_{10}(e)} \approx 2.303\,\log_{10}(A)$$

Toolkit A.7 Going to Extremes: When Values Tend to Zero or Infinity

It is helpful to consider what happens to values in our equations as a given value tends to extremes (*i.e.* tending to zero and tending to infinity). Let's use this approach to look at what happens in the Arrhenius equation as the temperature gets very low (*i.e.* $T \rightarrow 0$ K) and as the temperature gets very high (*i.e.* $T \rightarrow \infty$ K).

$T \rightarrow 0$ K	$T \rightarrow \infty$ K
$\dfrac{E_a}{RT} \rightarrow \infty$	$\rightarrow 0$

So, if $T = 0$ K, then the fraction E_a/RT evaluates to infinity; conversely, as $T \rightarrow \infty$ K, E_a/RT evaluates to zero. This just gives us another set of extreme values. What now is the effect on the exponential?

The 'right arrow' notation is used to mean 'approaches' or 'tends towards'. The representation '$T \rightarrow 0$ K' then means, 'As temperature, T, approaches zero kelvin.'

$T \rightarrow 0\,K$	$T \rightarrow \infty\,K$
$\dfrac{E_a}{RT} \rightarrow \infty$	$\rightarrow 0$
$e^{E_a/RT} \rightarrow e^{\infty} = \infty$	$\rightarrow e^0 = 1$

It looks like not much has changed, but now let's see what happens to the *decaying exponential*:

	$T \rightarrow 0\,K$	$T \rightarrow \infty\,K$
	$\dfrac{E_a}{RT} \rightarrow \infty$	$\rightarrow 0$
	$e^{E_a/RT} \rightarrow e^{\infty} = \infty$	$\rightarrow e^0 = 1$
$e^{-E_a/RT} \equiv \dfrac{1}{e^{E_a/RT}}$	$\rightarrow \dfrac{1}{\infty} = 0$	$\rightarrow \dfrac{1}{1} = 1$

We now have two distinct possibilities for a given reaction; when temperatures are very high, and when temperatures are very low.

At very low temperatures	At very high temperatures
$T \rightarrow 0\,K$	$T \rightarrow \infty\,K$
$k \rightarrow A \times 0$	$k \rightarrow A \times 1$

This means that our decaying exponential only varies between 0 and 1; in other words, it is a *probability*.

Toolkit A.8 The More Formal Way of Combining Errors of Many Functions

If we have a function a, which is dependent on a number of different values, $a(x, y, z...)$, and these different values are combined using a range of mathematical functions (any combination of addition and subtraction, multiplication and division, logarithms, powers and exponentials), then the uncertainty of a is given by[‡]

$$\partial a = \sqrt{\left(\frac{\partial a}{\partial x}\,\delta x\right)^2 + \left(\frac{\partial a}{\partial y}\,\delta y\right)^2 + \left(\frac{\partial a}{\partial z}\,\delta z\right)^2 \cdots} \qquad (A.2)$$

In plain English, we differentiate the function with respect to each term and multiply that result by the uncertainty in that term, then we take the square root of the sum of the squares of that value.

Practically this is demonstrated in the following example.

[‡]Take care not to confuse the symbol for a partial derivative, ∂ with that of the uncertainty, δ. This notation is potentially confusing, however it is what is used in many texts.

The heat provided in an isothermal expansion of a gas is given by eqn (A.3), where n is the number of moles of gas, T is the thermodynamic temperature, and V_f and V_i are the final and initial volumes, respectively:

$$q = nRT \ln\frac{V_f}{V_i} \qquad (A.3)$$

Therefore, assuming that each of the variable terms may have an error in it:

$$\delta q = \sqrt{\left(\frac{\partial q}{\partial n}\delta n\right)^2 + \left(\frac{\partial q}{\partial T}\delta T\right)^2 + \left(\frac{\partial q}{\partial V_f}\delta V_f\right)^2 + \left(\frac{\partial q}{\partial V_i}\delta V_i\right)^2} \qquad (A.4)$$

Differentiating each of these terms:

$$\delta q = \sqrt{\left(RT\ln\frac{V_f}{V_i}\delta n\right)^2 + \left(nR\ln\frac{V_f}{V_i}\delta T\right)^2 + \left(\frac{nRT}{V_f}\delta V_f\right)^2 + \left(\frac{nRT}{V_i}\delta V_i\right)^2}$$

$$(A.5)$$

which simplifies to:

$$\delta q = \sqrt{\left(\frac{q}{n}\delta n\right)^2 + \left(\frac{q}{T}\delta T\right)^2 + \left(\frac{nRT}{V_f}\delta V_f\right)^2 + \left(\frac{nRT}{V_i}\delta V_i\right)^2} \qquad (A.6)$$

This represents a rigorous approach to the analysis of uncertainties, however for most purposes in a teaching laboratory the approach shown in Chapter 3 will usually suffice. For help with differentiation, see Chapter 7.

Toolkit A.9 Geometric Relations

There are some fundamentals of geometry which we need to know in order to apply logic to our geometry problems. These are listed here.[§]

1. Axiom: Angles in a full rotation sum to 360° (2π radians).
2. Axiom: Angles on a straight line sum to 180° (π radians).
3. Angles in a triangle sum to 180° (π radians).
4. Angles in a polygon with n sides sum to $(180[n - 2])°$ $([n - 2]\pi$ radians).

[§]Some of these can be derived; others, known as *axioms*, are statements of truth which cannot be proved. They simply *are!* The axioms are labelled in this list.

Toolkit A.10 Pythagoras' Theorem

Pythagoras' theorem describes the relationship between sides on a right-angled triangle. For a generic right-angled triangle:

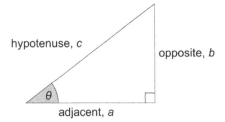

Pythagoras' theorem states: 'The square on the hypotenuse is equal to the sum of the squares on the other two sides.' Mathematically:

$$a^2 + b^2 = c^2$$

This relationship allows us to calculate the length of the remaining side of a right-angled triangle if we know any two sides. This finds use in this text in discussing trigonometry, considering vectors and the geometric visualisation of complex numbers.

Toolkit A.11 Factorial Numbers

A factorial is a way of notating a series multiplication of integers. It carries the notation of an exclamation mark, '!'. Simply, a factorial is notated as:

$$n! = n \times (n-1)!$$

so

$$n! = n \times (n-1)!$$
$$5! = 5 \times 4!$$
$$= 5 \times 4 \times 3!$$
$$= 5 \times 4 \times 3 \times 2!$$
$$= 5 \times 4 \times 3 \times 2 \times 1!$$
$$= 5 \times 4 \times 3 \times 2 \times 1 \times 0!$$
$$= 120$$

One thing which immediately jumps out is the number 0!. For various reasons, 0! evaluates to 1. This seems confusing at first; however, a discussion of the reasons for this result is outwith the scope of this text. To look at it simply, if $0! = 0$, then every factorial would evaluate to zero—a completely useless result!

When calculating probabilities, we may sometimes make use of a 'partial factorial'; e.g. we may have to calculate

$$M = 59 \times 58 \times 57 \times 56 \times 55 \times 54$$

This is obviously not equal to $59!$, so we will tend to notate these calculations in a different way

$$M = 59 \times 58 \times 57 \times 56 \times 55 \times 54$$
$$59! = 59 \times 58 \times 57 \times 56 \times 55 \times 54 \times 53!$$
$$= M \times 53!$$
$$M = \frac{59!}{53!}$$

This result is frequently used in probability calculations where large factorial chains can rapidly become unwieldy to manipulate.

Toolkit A.12 Differentiating Sine, Cosine and Exponential Functions

Sine, cosine and the exponential of Euler's number e can all be expressed in a similar manner using a Maclaurin series representation.

The expansion of the exponential of Euler's number is shown in eqn (A.7), with the expansions of sine and cosine shown in eqn (A.8 and A.9), respectively.

$$e^{Ax} = \frac{(Ax)^0}{0!} + \frac{(Ax)^1}{1!} + \frac{(Ax)^2}{2!} + \frac{(Ax)^3}{3!} + \cdots + \frac{(Ax)^\infty}{\infty!} \quad (A.7)$$

$$\sin(Bx) = \frac{(Bx)^1}{1!} - \frac{(Bx)^3}{3!} + \frac{(Bx)^5}{5!} - \frac{(Bx)^7}{7!} + \frac{(Bx)^9}{9!} - \cdots \quad (A.8)$$

$$\cos(Cx) = \frac{(Cx)^0}{0!} - \frac{(Cx)^2}{2!} + \frac{(Cx)^4}{4!} - \frac{(Cx)^6}{6!} + \frac{(Cx)^8}{8!} - \cdots \quad (A.9)$$

From these expansions, we can apply our principle of differentiation to each function in turn:

$$f(x) = e^{Ax} = \frac{(Ax)^0}{0!} + \frac{(Ax)^1}{1!} + \frac{(Ax)^2}{2!} + \frac{(Ax)^3}{3!} + \cdots + \frac{(Ax)^\infty}{\infty!}$$

$$= \frac{A^0 x^0}{0!} + \frac{A^1 x^1}{1!} + \frac{A^2 x^2}{2!} + \frac{A^3 x^3}{3!} + \cdots + \frac{A^\infty x^\infty}{\infty!}$$

$$f'(x) = \frac{0 \times A^0 x^{-1}}{0!} + \frac{1 \times A^1 x^0}{1!} + \frac{2 \times A^2 x^1}{2!} + \frac{3 \times A^3 x^2}{3!} + \cdots + \frac{A^\infty x^\infty}{\infty!}$$

$$= 0 + \frac{A \times A^0 x^0}{0!} + \frac{A \times A^1 x^1}{1!} + \frac{A \times A^2 x^2}{2!} + \cdots + \frac{A^\infty x^\infty}{\infty!}$$

$$= A\frac{A^0 x^0}{0!} + A\frac{A^1 x^1}{1!} + A\frac{A^2 x^2}{2!} + A\frac{A^3 x^3}{3!} + \cdots + A\frac{A^\infty x^\infty}{\infty!}$$

$$= Ae^{Ax}$$

$$g(x) = \sin(Bx) = \frac{(Bx)^1}{1!} - \frac{(Bx)^3}{3!} + \frac{(Bx)^5}{5!} - \frac{(Bx)^7}{7!} + \frac{(Bx)^9}{9!} - \cdots$$

$$= \frac{B^1 x^1}{1!} - \frac{B^3 x^3}{3!} + \frac{B^5 x^5}{5!} - \frac{B^7 x^7}{7!} + \frac{B^9 x^9}{9!} - \cdots$$

$$g'(x) = \frac{1 \times B^1 x^0}{1!} - \frac{3 \times B^3 x^2}{3!} + \frac{5 \times B^5 x^4}{5!} - \frac{7 \times B^7 x^6}{7!} + \frac{9 \times B^9 x^8}{9!} - \cdots$$

$$= B\frac{B^0 x^0}{0!} - B\frac{B^2 x^2}{2!} + B\frac{B^4 x^4}{4!} - B\frac{B^6 x^6}{6!} + B\frac{B^8 x^8}{8!} - \cdots$$

$$= B\cos(Bx)$$

$$h(x) = \cos(Cx) = \frac{(Cx)^0}{0!} - \frac{(Cx)^2}{2!} + \frac{(Cx)^4}{4!} - \frac{(Cx)^6}{6!} + \frac{(Cx)^8}{8!} - \cdots$$

$$= \frac{C^0 x^0}{0!} - \frac{C^2 x^2}{2!} + \frac{C^4 x^4}{4!} - \frac{C^6 x^6}{6!} + \frac{C^8 x^8}{8!} - \cdots$$

$$h'(x) = \frac{0 \times C^0 x^{-1}}{0!} - \frac{2 \times C^2 x^1}{2!} + \frac{4 \times C^4 x^3}{4!} - \frac{6 \times C^6 x^5}{6!} + \frac{8 \times C^8 x^7}{8!} - \cdots$$

$$= 0 - C\frac{C^1 x^1}{1!} + C\frac{C^3 x^3}{3!} - C\frac{C^5 x^5}{5!} + C\frac{C^7 x^7}{7!} - \cdots$$

$$= -C\sin(Cx)$$

Here we see how the index of the exponential remains unchanged through differentiation and how the overall value is scaled by the constant A, and likewise for the constants B and C in the differentiation of sine and cosine functions.

Toolkit A.13 Solving a Quadratic Equation

A quadratic equation has the general form

$$y = ax^2 + bx + c \qquad\qquad (A.10)$$

The terms a, b and c are constants and the variable in this instance is represented by x, though this can appear in many different guises in chemistry.

When plotted, these form the classic shape of a quadratic equation: the parabola.

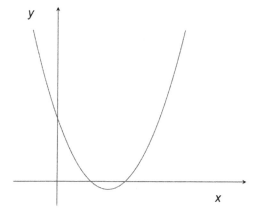

To 'solve' the quadratic, we are typically looking for the 'roots', *i.e.* the points at which the curve crosses the x-axis. We can do this in a number of ways, whether through factorising the equation, through completing the square or applying the quadratic formula.

Solving by Factorising

Consider the equation:

$$y = x^2 - 3x + 2$$

This can be factorised to find the roots—that is, when $y = 0$. We set the equation equal to zero, and then try to find which factors multiply together to give the original equation:

$$x^2 - 3x + 2 = (x - 1)(x - 2) = 0$$

From this, we can say that one of the factors must be zero (otherwise, we won't get a result of zero); therefore:

$$(x - 1) = 0 \qquad \text{or} \qquad (x - 2) = 0$$
$$x = +1, \qquad\qquad x = +2$$

Solving by Completing the Square

In this example, we assume that the solution is based on the square of a single factor. The equation is unlikely to be a square itself, so we add or subtract units to it to 'complete the square':

$$y = 16x^2 + 24x + 5$$

We start with a square number; $16x^2 = (4x)^2$. This is a good indication that we can use the technique of completing the square, so we try to find factors based on this. We can identify the factor as:

$$(4x + 3)^2$$

When we square this, however, we see that we obtain an equation which is not quite equal to the original equation:

$$(4x + 3)^2 = 16x^2 + 24x + 9$$

To make this the same as the original equation, we need to subtract the value of 4; therefore, we can now write our equation as:

$$y = 16x^2 + 24x + 5$$
$$= \left(16x^2 + 24x + 9\right) - 4$$
$$= \left(4x + 3\right)^2 - 4$$

We can now apply the same technique as before to solve when $y = 0$, remembering that a square root has two possible solutions:

$$\left(4x + 3\right)^2 - 4 = 0$$
$$\left(4x + 3\right)^2 + 4$$

$$(4x + 3)^2 = +2 \quad \text{or} \quad (4x + 3) = -2$$
$$4x = -1 \qquad\qquad 4x = -5$$
$$x = -\frac{1}{4} \qquad\qquad x = -\frac{5}{4}$$

Using the Quadratic Formula

The quadratic formula is often taught in schools and many will be able to remember its form:

$$x = \frac{-b \pm \sqrt{b^2 - 4ac}}{2a}$$

It offers a quick solution to solve any quadratic, as well as offering a means to identify solutions which have complex roots.

Its origin is based on completing the square and is simply a formulaic approach to this. The derivation of this equation from the process of 'completing the square' is outwith the scope of this text; however many derivations are available online.

Reading List

This text has been prepared largely from our experience of having taught and supported mathematics to first-year chemists over the past ten years; with this in mind, it is difficult to pinpoint exactly which texts inspired particular inclusions in the course. However, we have found the texts listed below to be very helpful in support of our teaching, and will be an excellent 'follow on' text from this.

- P. Atkins, J. de Paula and J. Keeler, *Physical Chemistry*, Oxford University Press, 11th edn, 2017.
- M. Cockett and G. Doggett, *Maths for Chemists*, Royal Society of Chemistry, 2nd edn, 2012.
- I. Hughes and T. P. A. Hase, *Measurements and their Uncertainties: A Practical Guide to Modern Error Analysis*, Oxford University Press, 2010.
- P. Monk and L. J. Munro, *Maths for Chemistry: A chemist's toolkit of calculations*, Oxford University Press, 2010.
- E. Steiner, *The Chemistry Maths Book*, Oxford University Press, 2008, 2nd edn.
- J. R. Taylor, *An Introduction to Error Analysis: The Study of Uncertainties in Physical Measurements*, University Science Books, USA, 2nd edn, 1997.

Index